Student Handbook and Solutions Manual

Essentials *of* Genetics

Second Edition

Student Handbook and Solutions Manual

Harry Nickla, Ph.D.
Creighton University

Essentials of Genetics

Second Edition

William S. Klug
Michael R. Cummings

Prentice Hall Upper Saddle River, NJ 07458

Production Editor: *Carole Suraci*
Production Supervisor: *Joan Eurell*
Acquisitions Editor: *Sheri Snavely*
Supplements Editor: *Mary Hornby*
Cover Design: *Kevin Kall/Paul Gourhan*
Cover Photo: *George Mattei*
Production Coordinator: *Ben Smith*

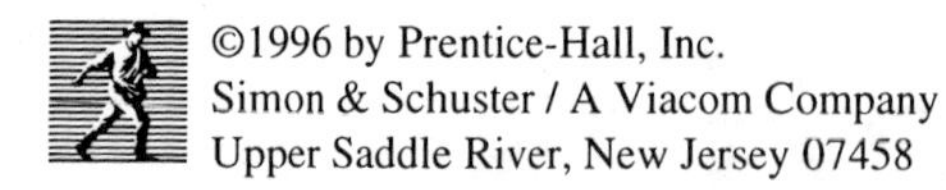

Simon & Schuster / A Viacom Company
Upper Saddle River, New Jersey 07458

Printed in the United States of America

10 9 8 7 6 5 4 3 2

ISBN: 0-13-381369-X

Prentice-Hall International (UK) Limited, *London*
Prentice-Hall of Australia Pty. Limited, *Sydney*
Prentice-Hall Canada Inc., *Toronto*
Prentice-Hall Hispanoamericana, S.A., *Mexico*
Prentice-Hall of India Private Limited, *New Delhi*
Prentice-Hall of Japan, Inc., *Tokyo*
Simon & Schuster Asia Pte. Ltd., *Singapore*
Editora Prentice-Hall do Brasil, Ltda., *Rio de Janeiro*

Contents

Introduction

Purpose of this book

The intent of this book is to help you understand introductory genetics as presented in **Essentials of Genetics**. To do so, you must build a conceptual framework in which to place various experiments, examples, and illustrations that are prevalent in genetics and which usually represent the basis of specific homework problems and test questions. To succeed, you must be able to recognize from where in that framework each particular test question is drawn, and what examples are likely to pertain to each concept.

A first course in genetics can be a humbling experience for many students. It is possible that the lowest grades received in one's major, or even in one's undergraduate career, may be in genetics. It is not unusual for some students to become frustrated with their own inability to succeed in genetics. This frustration is felt by most teachers as they field most of the following student comments:

Student frustrations

"I studied all the material but failed your test."

"I must have a mental block to it. I just don't get it. I just don't understand what you are asking."

"I know all the material, but I can't take your tests."

"Where did you get that question? I didn't see anything like that in the book or in my notes."

"This is the first test I have **ever** failed."

"I helped three of my friends last night and I got the lowest grade."

"I am getting a "D" in your course and I have never received less than a "B" in my whole life."

"I stayed up all night studying for your exam and I still failed."

Similar to Algebra

Think back to the first time you encountered "word problems" in your first algebra class. How many times did you say to yourself, your parents, or to your teacher,

> "I hate word problems, I just can't understand them, and why do I need to learn this anyway, I'll never use it."

At that time you had two choices, drop out and be afraid of problem-solving for the rest of your life (which unfortunately happens too often) or regroup, seek help, strip away distractions, and focus in on learning something new and powerful. Because you are taking genetics, you probably succeeded in algebra, perhaps with difficulty at first, and you will probably succeed in genetics.

You were forced in algebra to convert something real and dynamic (two trains leaving at different times from different stations at different speeds - when do they meet?) to a somewhat abstract formula which can be applied to an infinite

number of similar problems. In genetics you will again learn something new. It will involve the conversion of something real and dynamic (genes, chromosomes, hereditary elements, gamete formation, gene splicing, and evolution) to an array of general concepts (similar to mathematical formulas) which will allow you to predict the outcome of an infinite number of presently known and yet to be discovered phenomena relating to the origin and maintenance of life.

Mental Pictures and Symbols

When working almost any algebra word problem, it is often helpful to make a simple drawing which relates, in space, the primary participants. From that drawing one can often predict or estimate a likely outcome. A mathematical formula and its solution provides the precise outcome. In understanding genetics it is often helpful to make drawings of the participants whether they be crosses (*Aa* X *Aa*), gametes (*A* or *a*), or the interactions of molecules (anticodon with codon).

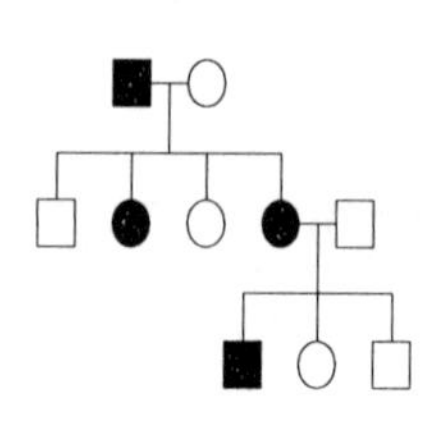

As with algebra, the symbolism used to represent a multitude of structures, movements, and interactions, is abstract, informative, and fundamental to understanding the discipline. It is the set of symbols and the relationships among symbols which comprise the universal set of concepts (paradigms) which, as specific examples, make up the framework of genetics. Test questions and problems which, as examples, may be completely unfamiliar to the student, nevertheless refer directly to the basic concepts of genetics.

Processes and Concepts

- ***Cytological Basis: Mitosis and Meiosis***
- ***Mendelian Genetics, Modifications***
- ***Quantitative Genetics***
- ***Linkage and Mapping***
- ***Non-Mendalian Inheritance***
- ***Variation in Chromosome Number and Arrangement***
- ***DNA: Replication, Storage, and Expression***
- ***Organization of Genes and Chromosomes***
- ***Mutation***
- ***Proteins***
- ***Bacterial and Viral Genetics***
- ***Recombinant DNA Technology***
- ***Regulation and Development, Behavior***
- ***Applications: Immunity and Cancer***
- ***Population and Evolutionary Genetics***

A Student's Relationship to the Genetics Instructor

Teachers usually adjust the level of a course, the selection of a text, test questions, and lecture material, on two criteria:

(1) the capability of the students, which is determined by factors such as the course prerequisite pattern, the entry standards of the institution, and

(2) the experience and expertise of the instructor.

While the instructor may do his/her best to present the material clearly, the burden is on the student to learn genetics. Regardless of how hard a teacher tries to explain certain concepts, the student must be an active participant in the learning process. A student can not enter a genetics classroom, forgetting all that was learned in the prerequisite course(s), and expect the instructor to start from scratch and teach them genetics.

There are interesting, important, and somewhat complex concepts to learn. Such learning requires a focused effort not only by the instructor but also by the student. The student should expect that a competent instructor is attempting to teach them how life, with its variation and constancy, is passed from one generation to the next. The instructor expects that students are responding with a thoughtful, mature, disciplined effort to learn. Anything short of those expectations is likely to result in disappointment on the part of the instructor and the student.

Granted, there may be idiosyncrasies of a given instructor which may be distracting or even annoying to students. But mature students are able to dismiss surface distractions and focus on the subject material. They are unwilling to let an individual stand between them and their right to an understanding of significant biological information.

The Tests are Fair

Because many students may fail a particular test, it is no indication that the test is not fair. It is expected that some students will have a more difficult time with certain concept areas than others and as a result will do poorly on certain tests. It would be a tremendous disservice to our students if instructors "watered down" or omitted difficult material so that more students would receive better grades. For a variety of reasons, there are temptations (student evaluations, fewer enemies) to give high grades and fail few.

Experienced teachers recognize conceptual areas which are the most difficult for students and often attempt different teaching strategies or hold extra review sessions for students when those areas are encountered. It is up to the student to take advantage of the instructor's offer to help.

There is considerable uniformity in the instruction of genetics. Instructors expect students to learn certain fundamental concepts before passing grades are awarded. Students should expect little compromise in this aspect of evaluation.

It is likely that a student will encounter a test question which involves an unfamiliar example or situation. Genetics teachers may use literally thousands of ways to test a student's understanding of a simple dihybrid cross. Students who do well in genetics are able to focus on understanding concepts, rather than memorizing a multitude of examples.

Attendance and Attention are Mandatory

Many professors do not take attendance in lectures, therefore, it is likely that some students will opt to take a day off now and then. Unless those students are excellent readers and excellent students in general, continual absences will ususally result in failure.

Remember how difficult it was to set-up and understand the first algebra word problem on your own. It is likely that your ultimate source of understanding came from the course instructor. While using the text is important in your understanding of genetics, the teacher can walk you through the concepts and strategies much more efficiently than a text because a text is organized in a *sequential* manner. A good teacher can "cut and paste" an idea from here and there as needed. To benefit from the wisdom of the instructor, the student must concentrate during the lecture session rather than sit passively taking notes, assuming that the ideas can be figured out at a later date. Too often the student will not be able to relate to notes passively taken weeks before.

It is necessary for students to attend class and take advantage of the instructor's insights during the lecture sessions. The instructor will not be able to cover all the material in the text. Parts will be emphasized while other areas may be omitted entirely. Since it is the instructor who writes and grades the tests, who is in a better position to prepare the students for those tests?

There is no magic formula for understanding genetics or any other discipline of significance. Learning anything, especially at the college level, requires time, patience, and confidence. First, a student must be willing to focus on the subject matter for an hour or so each day over the entire semester (quarter, trimester, etc.). Study time must be free of distractions and pressured by the presence of clear, realistic goals.

The student must be patient and disciplined. It will be necessary to study when there are no assignments due and no tests looming.

The majority of successful students are willing to read the text ahead of the lecture material, spend time thinking about the concepts and examples, and work as many sample problems as possible. They study for a period of time, stop, then return to review the most difficult areas. They do not try to cram information into marathon study sessions a few nights before the examinations. While they may get away with that practice on occasion, more often than not, understanding the concepts in genetics requires more mature study habits and preparation.

Perhaps a Different Way of Thinking

Because the acquisition of problem-solving ability requires that students rely on new and important ways of seeing things rather than memorizing the book and notes, some students find the transition more difficult than others. Some students are more able to deal in the abstract, concept-oriented framework than others. Students who have typically relied on "pure memory" for their success, will find a need to focus on concepts and problem-solving. They may struggle at first just as they may have struggled with the first word problem in algebra. But the reward for such struggle is intellectual growth.

Students should expect to grow intellectually in a genetics course and with such growth will come an increased ability to solve a variety of problems beyond genetics. Problem-solving is a process, a style, which can be applied to many disciplines. Success comes from probing deeply in a few areas to see how problems are approached in a given discipline. Then, because problems are usually solved in a fairly consistent manner, a given problem-solving approach can often be applied to a variety of activities.

At each particular point in a lifetime, one has different capabilities. Certain intellectual areas are less developed than others, perhaps because of a particularly simulating teacher in grade school or a particularly poor teacher. It may be a neighbor or a relative that gave you, by chance or by insight, the stimulation (in the form of a particular book or an extended conversation or explanation) that helped you take a mental leap. There is often the impression that if you work hard enough, you can accomplish anything. That is simply not true in all one's endeavors. By working hard in a focused, concentrated fashion, one can accomplish a great many things, perhaps the most important of which is the confidence that comes from achieving goals, even if small at first. Hard work, in combination with intellectual maturity, often leads to academic success.

How to Study

Genetics is a science which involves symbols (*A*, *b*, *p*), structures (chromosomes, ribosomes, plasmids), and processes (meiosis, replication, translation) which interact in a variety of ways. Models describe the manner in which hereditary units are made, how they function, and how they are transmitted from parent to offspring. Because many parts of the models interact both in time and space, genetics can not be viewed as a discipline filled with facts which should be memorized. Rather, one must be, or become, comfortable with seeking to understand not only the components of the models but also how the models work.

Time,
Work,
Patience

One can memorize the names and shapes of all the parts of an automobile engine, but without studying the interrelationships among the parts in time and space, one will have little understanding of the real nature of the engine. It takes time, work, and patience to see how an engine works, and it will take time, work, and patience to understand genetics. And just as one is likely to be fascinated with the movement and power of an engine, such fascination is quite likely to come with an understanding of genetics.

Don't cram. A successful tennis player doesn't learn to play tennis overnight; therefore, you can't expect to learn genetics under the pressure of night-long cramming. It will be necessary for you to develop and follow a realistic study schedule for genetics as well as the other courses you are taking. It is important that you focus your study periods into intensive, but relatively short sessions each day throughout the entire semester (quarter, trimester). Because genetics tests often require you to think "on the spot" it is very important that you get a good night's sleep before each test. Avoid caffeine in the evening before the test because a clear, rested, well-prepared mind will be required.

Study when there are no tests

Develop a realistic schedule

Tuesday

Study times:	Subject:
1st hour:	Genetics
	History
	Physics
	Literature

Monday

Study times:	Subject:
1st hour:	Psychology
2nd hour:	Genetics
Recreation	
3rd hour:	History
4th hour:	Literature

Study goals. The instruction of genetics is often divided into large conceptual units. A test usually follows each unit. It will be necessary for you to study genetics on a routine basis long before each test. To do so, set specific study goals. Adhere to these goals and don't let examinations in one course interfere with the study goals of another course. Notice that each course being taken is handled in the same way — study ahead of time and don't cram.

Develop a monthly plan (Example)

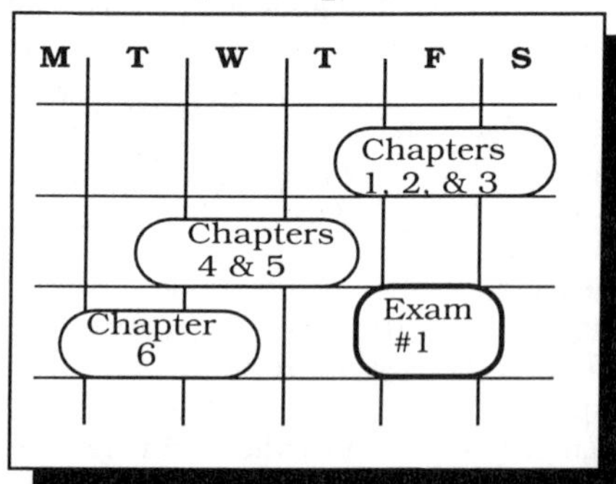

Read ahead. You have been told that it is important to read the assigned material before attending lectures. This allows you to make full use of the information provided in the lecture and to concentrate on those areas which are unclear in the readings. An opportunity is often provided for asking questions. Your questions will be received much more favorably if you can say that after reading the book and listening to the lecture a particular point is still unclear. It is very likely that your question will be quickly dealt with to your benefit and the benefit of others in the class.

Develop a plan for the semester (Example)

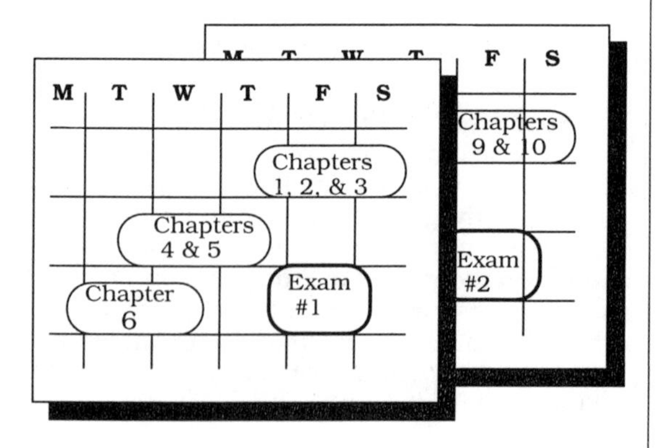

Work the assigned problems. The basic concepts (or models) of genetics are really quite straightforward but there are many examples which apply to these concepts or models. To help students adjust to the variety of examples and approaches to various concepts, instructors often assign practice problems from the back of each chapter. If your instructor has assigned certain problems, finish working them *at least* one week before each examination. Before starting with a set of problems, read the chapter carefully and consider the information presented in class.

Suggestions for working problems:

(1) work the problem *without* looking at the answer

(2) check your answer in this book

(3) if incorrect, work the problem again

(4) if still incorrect, you don't understand the concept

(5) re-read your lecture notes and the text,

(6) work the problem again

(7) if you still don't understand the solution, mark it, and go to the next problem.

(8) In your next study session, return to those problems which you have marked. Expect to make mistakes and learn from those mistakes. Sometimes what is difficult to see one day may be obvious the next day. If you are still having problems with a section, schedule a meeting with your instructor. Usually the problem can be cleared up in a few minutes.

You will notice that in this book, I have presented the solution to each problem. I provide different ways of looking at some of the problems. Instructors often take a problem directly from those at the end of the chapters or they will modify an existing problem. Reversing the "direction" of a problem is a common approach. Instead of giving characteristics of the parents and asking for characteristics of the offspring, the question may provide characteristics of the offspring and ask for particulars on the parents.

Separate examples from concepts. As mentioned earlier, genetics boils down to a few (perhaps 15 to 20) basic concepts; however there are many examples which apply to those concepts. Too often students have trouble separating examples from the concepts. Notice that in the "Sample Test" sections in this book, I have made such separations clear. Examples allow you to picture, in concrete terms, various phenomena but they don't exemplify each phenomenon or concept in its entirety.

Caution in the use of old examinations. Often it is customary for students to request or otherwise obtain old examinations from previous students. Such a practice is loaded with pitfalls. First, students often, albeit unconsciously, find themselves "second guessing" about questions on an upcoming examination. They forget that an examination usually only tests over a subset of the available information in a section. Therefore entire "conceptual areas" may be available which have not appeared on recent exams.

Often the reproductions of old examinations are of poor quality (having been copied and passed around repeatedly) and it is difficult to determine whether the answer provided is the correct one or if it was incorrect and marked wrong. In addition, if a question has the same general structure as one on a previous examination, but is modified, students often provide an answer for the "old" question rather than the one being asked.

Granted, it is of value to see the format of each question and the general emphasis of previous examinations, but remember that each examination is potentially a new production capable of covering areas which have not been tested before. This is especially likely in a course such as genetics where the material changes very rapidly. Don't try to figure out what will be asked. Study all the material as well as possible.

Structure of this Book

The intent of this book is to help you understand the concepts of genetics as given in the text and most likely in the lectures, then to apply these concepts to the solution of all problems and questions at the ends of the chapters. Rather than merely provide you with the solutions to the problems, I have tried to walk you through each component of each question so that you can see where information is obtained and how it can be applied in the solution.

Vocabulary: organization and listing of terms. Understanding the vocabulary of a discipline is essential to understanding the discipline. Throughout the *Essentials* text you will find terms in bold print. Such terms generally refer to structures or substances, processes/methods, and concepts. I have separated these terms and *other important terms* into these categories.

Structures and Substances

Processes/Methods

Concepts

Those terms or concepts which require special explanation or are more complex or intimately related to other terms are denoted with a code (F2.1, F21.2, etc.) which refers you to the figures immediately following the ***Concepts*** section.

You may see a term present in several categories or in categories other than those you would envision. Since categorization *per se* is of little significance, don't worry about category location. Use the listings as checklists to make certain that you understand the meaning of each term in each chapter. Also, by a given term's category, you can begin to understand whether it refers to a structure or substance, a process or method, or a more general concept. Notice that the various terms are not redefined. It is important that you use the *Essentials* text for the original definitions.

Concepts. In the section *"Vocabulary: Organization and Listing of Terms"* you will find a section called *"Concepts"* after which there may be a simple sketch or two to help you focus a particular concept. Such sketches are oversimplifications and you should fill in the details by examining the textbook and the lecture notes.

Understand the words and phrases of the discipline

Solved problems. Each of the problems at the end of each chapter is solved from a beginner's point of view. There are other features of this section. Many of the answers to the questions and problems will refer you to specific sections, usually specific tables and figures, of the *Essentials* text (*Essentials*-Fig.2.7, for example). Be certain that you fully understand the solution to each of the questions suggested or assigned by your instructor. Figures or tables referred to in this handbook are labeled as T or F, for example, T2.1 or F3.2

Supplemental questions. A series of solved sample test questions supplement the questions provided in the text and help you determine your level of preparation. These sample test questions are located at the end of each of the major conceptual units: Chapters 1-8, 9-14, and 15-22 (although your instructor may test over different chapter groups). Concepts relating to each question as well as common errors are presented in boxes before and after each answer.

Supplemental Questions

Concepts

Comprehensive Solution

Common errors

1

An Introduction to Genetics

Vocabulary: Organization and Listing of Terms

Historical

- Prehistoric times
  - domesticated animals
  - cultivated plants
- Greek Influence
  - Hippocrates
    - "humors"
  - Aristotle
    - vital heat
- Lamarck
  - pangenesis
  - acquired characteristics
    - use and disuse
- Epigenesis
- Preformation
  - cell theory
  - homunculus
    - ovists
    - spermists
  - spontaneous generation
- Fixity of species
  - Linnaeus
  - Kolreuter
    - hybridization
    - parental types
    - backcross
  - *Naturphilosophie*
- Evolution
  - Darwin (1859)
    - *The Origin of Species*
    - natural selection
    - pangenesis
- Genetics
  - Mendel (1865)
  - Rediscovery (1900)
    - Correns, de Vries, Von Tschermak
    - Bateson

Structures and Substances

- Center for heredity
  - nucleus
  - nucleoid region
  - viral head
- Genetic material
  - nucleotides
  - DNA (deoxyribonucleic acid)
    - double helix
      - information storage
      - replication
      - mutation
      - expression
  - RNA (ribonucleic acid)
    - nitrogenous bases
  - genes
  - chromosomes
    - chromatin fibers
    - centromere
- Associated substances/structures
  - amino acids
  - messenger RNA
  - ribosome
  - ribosomal RNA
  - transfer RNA
  - protein
    - enzymes

Processes/Methods

- Mitosis
- Meiosis
- Activation energy
- Transmission genetics
  - pedigree analysis
  - cytological investigations
  - chromosome theory of inheritance
  - karyotypes
- Molecular and biochemical analysis
  - recombinant DNA studies
- Genetic structure of populations
- Basic research
- Applied research
  - eugenics
    - Galton
      - positive, negative
  - euphenics
  - agriculture
    - "Green Revolution"
    - Borlaug
  - medicine
    - genetic counseling
    - human genetic engineering
    - immunogenetics

Concepts

- Genetic information
  - stored
  - altered
  - expressed
  - regulated
- Genetics
  - heredity
  - variation
    - alleles
    - phenotype
- Diploid number (2n) (F1.1)
  - polyploid
  - haploid number (F1.1)
  - homologous chromosomes (F1.2)
  - locus, loci
- Genetic variation
  - gene mutations
  - chromosomal aberrations (mutations)
- Genetic information
  - genetic code
  - transcription
  - translated (translation)

F1.1. Below is pictured the union of two gametes, each with the *n* chromosome number. The haploid chromosome number (n) is usually found in gametes while the diploid chromosome number (2n) is usually found in the zygote and somatic tissues of the individual. There are higher levels of "ploidy" such as 3n, 4n, and so on.

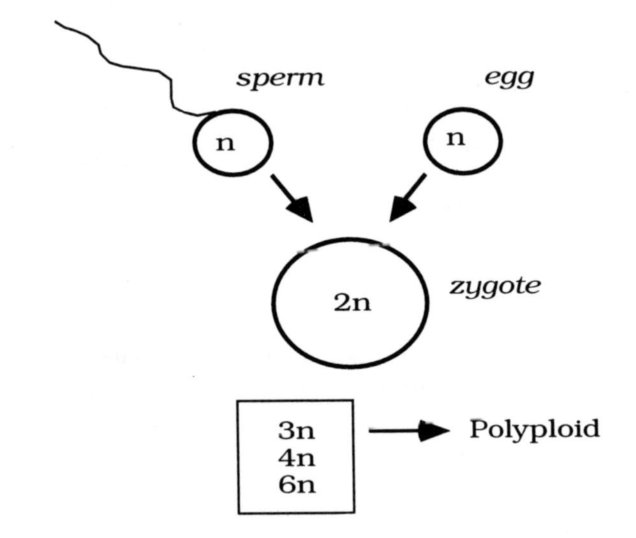

F1.2. One of the most important concepts to be mastered involves an understanding of homologous chromosomes. Each parent normally contributes one of each type of chromosome to the zygote. Since there are two gametes which unite to form the zygote, there must be two of each type of chromosome in each zygote. These two chromosomes of the same type are called homologous chromosomes or homologous pairs of chromosomes. Humans have 23 pairs of homologous chromosomes. One of each pair came from our mother, the other member of each pair came from our father (as shown in F1.1).

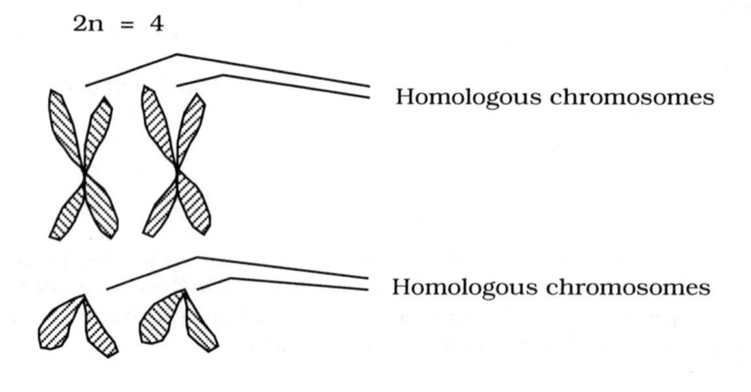

Criteria:

1. Size
2. Centromere position
3. Other similarities

Homologous chromosomes are similar in that they are generally of the same size, have the same centromere position, and have many other characteristics in common.

Solutions to Problems and Discussion Questions

1. Both were concerned with subjects of the reproduction, heredity, the origin of humans, and the shifting of interest from religious mythology to philosophical and scientific inquiries. Hippocrates argued that male semen is formed in various parts of the body (healthy or diseased) and transported through blood vessels to the testicles. Such "humors" carried the hereditary traits. Thus the theory of pangenesis was formed. Aristotle was critical of pangenesis because it did not explain the appearance of features which skipped generations. Aristotle suggested that semen contained a vital heat which could produce offspring in the form of the parents.

2. *Pangenesis* refers to a theory that various parts of the body contain "humors" which bear the hereditary traits and gather in the reproductive organs. *Epigenesis* refers to the theory that organisms are derived from the assembly and reorganization of substances in the egg which eventually lead to the development of the adult. *Preformationism* is a 17th century theory which states that the sex cells (eggs or sperm) contain miniature adults, called homunculi, which grow in size to become the adult. Each postulates a fundamental differnce in the manner in which organisms develop from hereditary determiners.

3. Darwin was aware of the physical and physiological diversity of members within and among various species. He was aware that varieties of organisms could be developed through selective breeding (domestication), that a species is not a fixed entity, and that while certain groups of organisms could be hybridized, other groups could not. He was aware of conflicts between religious views and the fossil record. He understood geology, geography, and biology and that organisms tend to leave more offspring than the environment can support.

4. Darwin's theory of natural selection proposed that more offspring are produced than can survive, and that in the competition for survival, those with favorable variations survive. Over many generations, this will produce a change in the genetic make-up of populations if the favorable variations are inherited. Darwin did not understand the nature of heredity and variation which, led him to lean toward older theories of pangenesis and inheritance of acquired characteristics.

5. Genes, linear arrays of nucleotides, usually exert their influence by producing proteins through the process of transcription and translation. Genes are the functional units of heredity. They associate, sometimes with proteins, to form chromosomes. During the cell cycle, chromosomes and therefore genes, are duplicated by a variety of enzymes so that daughter cells inherit copies of the parental hereditary information. Genes of eukaryotes are composed of DNA. A simplified diagram of the role of DNA in producing proteins is given below:

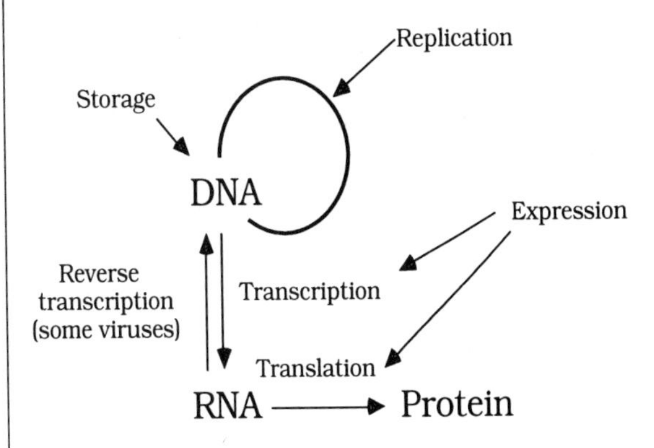

A simplified diagram of chromosomes is presented in F1.2.

6. *Transmission* genetics is the most classical approach in which the patterns of inheritance are studied through selective matings or the results of natural matings. Mendel observed results from precisely-defined matings and provided models based on transmission genetics.

A second approach involves physical, traditionally microscopic (light and electron), examination of chromosomes. With the discovery of mitotic and meiotic processes, and the knowledge that genes are located on chromosomes, much interest centers on the *cytological investigation* of chromosomes.

Molecular and biochemical analysis of the genetic material has recently evolved into one of the most exciting and rapidly growing subdisciplines of genetics. Originating in the early 1940s with studies of bacteria and viruses, much information as to the nature of gene expression, regulation, and replication has been provided. *Recombinant DNA technology* has had a significant impact in this area as well as others.

In *population* genetics, the interest is in the behavior of genes in groups of organisms (populations) often with an interest in the factors which change gene frequencies in time and space. Hence population geneticists are often interested in the process of evolution.

7. *Basic* research involves the study of the fundamental mechanisms of genetics as described in the answer to question #6 above. *Applied* research makes use of the information provided by basic research. Many applications in agriculture and medicine are described in the text.

8. Norman Borlaug applied Mendelian principles of hybridization and trait selection to the development of superior varieties of wheat. Such varieties are now grown in many countries, including Mexico, and have helped maintain the world supply of food. This change in worldwide agricultural food production has been called the "Green Revolution."

9. *Positive eugenics* encouraged parents displaying favorable characteristics to have large families while *negative eugenics* attempted to restrict reproduction for parents displaying unfavorable characteristics. *Euphenics* refers to medical genetic intervention designed to reduce the impact of defective genes on individuals.

10. In the last 40 years human transmission, cytological and molecular genetics have provided an understanding of many aspects of both plant and animal biology. In addition, much has been learned about many human diseases. There is promise that a certain amount of human suffering will be minimized by the application of genetics to crop production (disease resistance, protein content, growth conditions) and medicine. Major medical areas of activity include genetic counseling, gene mapping and identification, disease diagnosis, and genetic engineering.

Cell Division and Chromosomes

Vocabulary: Organization and Listing of Terms

Structures and Substances

- Cells
  - plasma membrane
    - glycoproteins
  - cell wall
    - cellulose
    - peptidoglycan
  - capsule
  - cell coat
    - AB antigens
    - MN antigens
    - receptor molecules
    - histocompatibility antigens
  - nucleus, nucleoid
    - genetic material
    - chromatin
    - chromosomes
      - histones
      - folded fiber model
      - nucleolus
      - nucleolar organizer (NOR)
      - rRNA
  - cytoplasm
    - colloidal
    - cytosol
      - cytoskeleton
        - microtubules, microfilaments
    - endoplasmic reticulum
    - ribosomes
    - mitochondria
    - chloroplasts
    - basal body
    - centrioles
    - spindle fibers
- Chromosomes
  - centromere, *p* arm, *q* arm
  - metacentric
  - submetacentric
  - acrocentric
  - telocentric

karyotype

sex-determining chromosomes

Mitosis

zygotes

centrosome, centriole

spindle fibers

chromatid

centromere, kinetochore

sister chromatids

cell plate

Meiosis

bivalent
tetrad
dyad
monad

chiasma (chiasmata)
synaptonemal complex

Gametogenesis

testes

spermatogonium
primary spermatocyte
secondary spermatocyte
spermatid
spermatozoa (sperm)

ovary

oogonium
primary oocyte
secondary oocyte
first polar body
ootid
second polar body
ova (ovum)

Processes/Methods

Mitosis

cell cycle

cytokinesis

interphase

S phase

G_1 and G_2

M phase

prophase

prometaphase

metaphase

anaphase

telophase

Regulation

significance

G_0

R point

cdc2, kinase

cyclins

p53

tumor suppressor gene

cell suicide

Meiosis

reductional

equational

- prophase I
    - leptonema
    - zygonema(synapsis)
    - pachynema
        - crossing over
    - diplonema
    - diakinesis (terminalization)
- metaphase I
- anaphase I
- telophase I
- prophase II
- metaphase II
- anaphase II
- telophase II

Spermatogenesis

Spermiogenesis

Oogenesis

- first meiotic division (at ovulation)
- second meiotic division (at fertilization)

Fertilization

Sexual reproduction

- reshuffles chromosomes
- provides for crossing over

Sporophyte

Gametophyte

Polytene chromosomes

- chromomeres
    - autoradiography
- puffs

Lampbrush chromosomes

- linear axis
- lateral loops

Concepts

Homologous chromosomes (F2.2)

- diploid number (2n) (F2.1)
- biparental inheritance
- loci (F2.2)
- haploid genome (haploid number, n) (F2.2)
- alleles (F2.2)

Mitosis (T2.1)

- identical daughters
- equivalent genetic information

Meiosis (F2.3)

- segregation
- independent assortment
- produces gametes or spores
- reshuffles genetic combinations (chromosomes)
- genetic recombination (crossing over)
- constant amount of genetic material
- production of variation

Fertilization

- reconstitution of genetic material

F2.1. Diagram showing the relationships among stages of interphase chromosomes, chromosome number, and chromosome structure in an organism with a diploid chromosome number of 4 (2n = 4). There are two pairs of chromosomes, one large metacentric, one smaller metacentric. Individual chromosomes can not be seen at interphase, therefore, the drawings represent chromosomes *as if* they could be individually viewed.

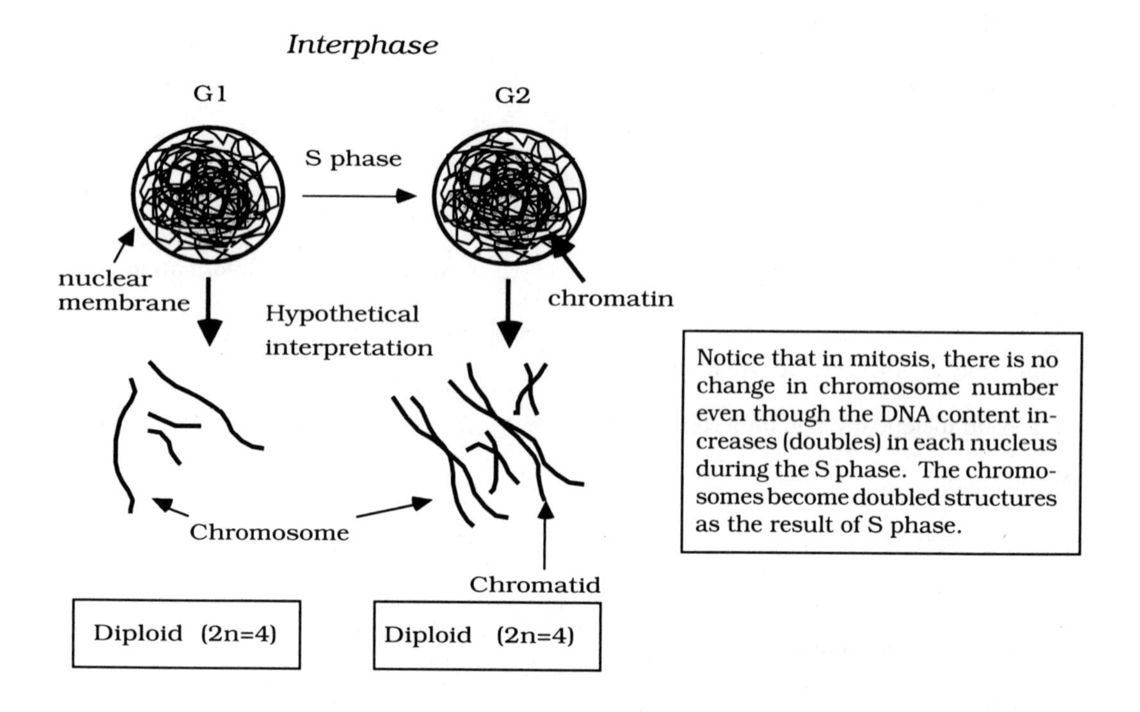

F2.2. Important nomenclature referring to chromosomes and genes in an organism where the diploid chromosome number is 4 (2n = 4). There are two pairs of chromosomes, one large metacentric, and one smaller metacentric. Sister chromatids are *identical* to each other while homologous chromosomes are *similar* to each other in terms of overall size, centromere position, function, and other factors as described in your text.

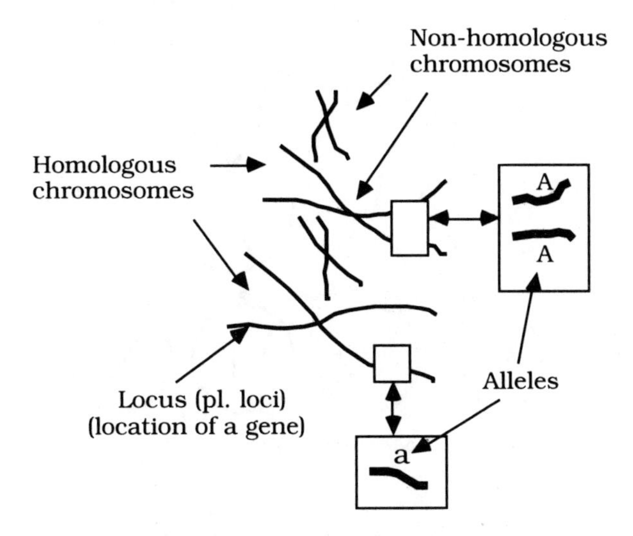

T2.1. Illustration of the relationship between chromosome number and stages of the mitotic cycle.

Cell cycle stage	***Chromosome number***	
	humans	**fruit flies**
Interphase	2n = 46	2n = 8
G1	2n = 46	2n = 8
S	2n - 46	2n = 8
G2	2n = 46	2n = 8
Mitosis		
Prophase	2n = 46	2n = 8
Metaphase	2n = 46	2n = 8
Anaphase	2n = 46	2n = 8
Telophase	2n = 46	2n = 8

F2.3. Illustrations of chromosomes of meiotic and mitotic cells in an organism with a chromosome number of 4 (2n = 4).

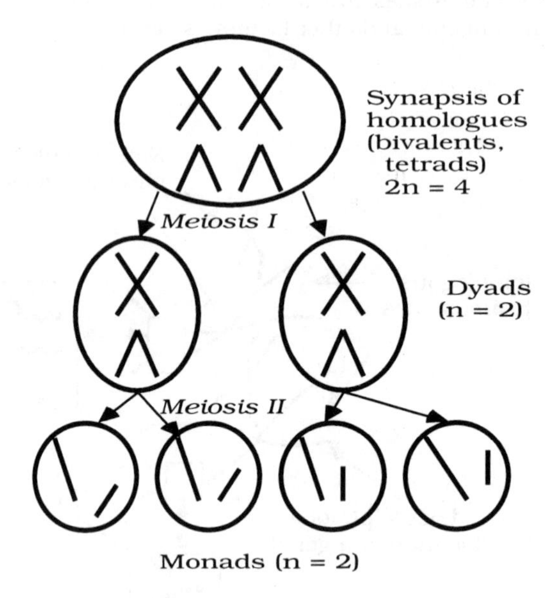

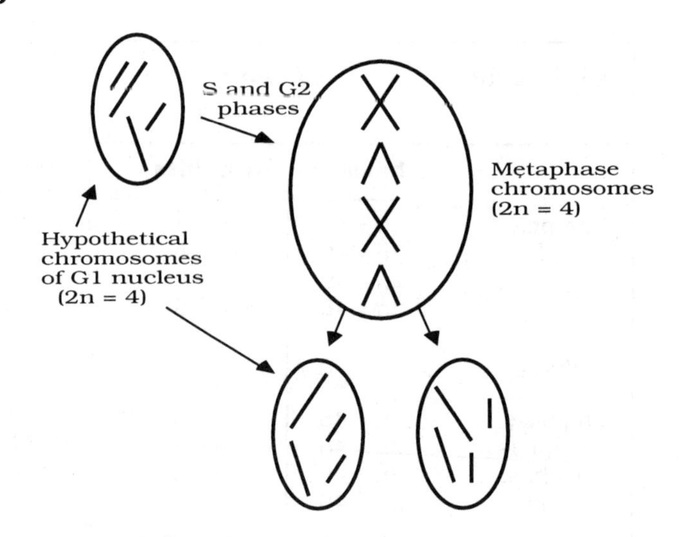

Solutions to Problems and Discussion Questions

1.

(a) During interphase of the cell cycle (mitotic and meiotic), chromosomes are not condensed and are in a genetically active, spread out, form. In this condition, chromosomes are not visible as individual structures under the microscope (light or electron). See F2.1 for a sketch of what *chromatin* might look like. Chromatin contains the genetic material which is responsible for maintaining hereditary information (from one cell to daughter cells and from one generation to the next) and production of the phenotype.

(b) The *nucleolus* (*pl. nucleoli*) is a structure which is produced by activity of the nucleolar organizer region in eukaryotes. Composed of ribosomal RNA and protein, it is the site for the production of ribosomes. Some nuclei have more than one *nucleolus*. Nucleoli are not present during mitosis or meiosis because in the condensed state of chromosomes, there is little or no RNA synthesis.

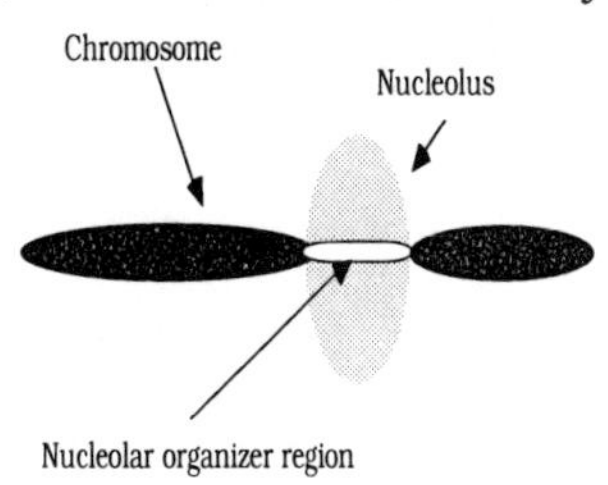

(c) The *ribosome* is the site where various RNAs, enzymes, and other molecular species assemble the primary sequence of a protein. That is, amino acids are placed in order as specified by messenger RNA. Ribosomes are relatively non-specific in that virtually any ribosome can be used in the translation of any mRNA. The structure and function of the ribosome will be described in greater detail in later chapters of the *Essentials* text.

(d) The *mitochondrion* (*pl. mitochondria*) is a membrane-bound structure located in the cytoplasm of eukaryotic cells. It is the site of oxidative phosphorylation and production of relatively large amounts of ATP. It is the trapping of energy in ATP which drives many of important metabolic processes in living systems.

(e) The *centriole* is a cytoplasmic structure involved (through the formation of spindle fibers) in the migration of chromosomes during mitosis and meiosis.

f) The *centromere* serves as an attachment point for sister chromatids (see F2.1), a point where spindle fibers attach to chromosomes. The centromere divides during mitosis and meiosis II, thus aiding in the partitioning of chromosomal material to daughter cells. Failure of centromeres or spindle fibers to function properly may result in nondisjunction.

2. One of the most important concepts to be gained from this chapter is the relationship which exists among chromosomes in a single cell. Chromosomes which are homologous share many properties including:

overall length; look carefully at F2.1 and F2.2 to see that each cell prior to anaphase I contains two chromosomes of approximately the same overall length.

position of the centromere (metacentric, submetacentric, acrocentric, telocentric); Again, look carefully at F2.1 and F2.2, and the *Essentials* text. Notice that in each if there is one metacentric chromosome, there will be another metacentric chromosome.

banding patterns; Look carefully at the *Essentials* text and notice how similar the banding patterns are of the homologous (side by side) chromosomes.

Notice also that sister chromatids have identical banding patterns as would be expected since sister chromatids are, with the exception of mutation, identical copies of each other. We would expect that homologous chromosomes would have banding patterns which are very similar (but not identical) because homologous chromosomes are genetically similar but not genetically identical.

type and location of genes; Notice in F2.2 that a *locus* signifies the location of a gene along a chromosome. What that really means is that for each characteristic specified by a gene, like blood type, eye color, skin pigmentation, there are genes located along chromosomes. The *order* of such loci is identical in homologous chromosomes, but the genes themselves, while being in the same order, may not be identical.

Look carefully at the inset (box) in the upper right portion of F2.2 and see that there are alternative forms of genes, *A* and *a*, at the same location along the chromosome. *A* and *a* are located at the same place and specify the same *characteristic* (eye color for example) but there are slightly different manifestations of eye color (*brown* vs. *blue* for example). Just as an individual may inherit gene *A* from the father and gene *a* from the mother, each zygote inherits one homologue of each pair from the father and one homologue of each pair from the mother.

autoradiographic pattern; homologous chromosomes tend to replicate during the same time of S phase.

Diploidy is a term often used in conjunction with the symbol *2n*. It means that both members of a homologous pair of chromosomes are present. Refer to F2.1 in this book. Notice that during mitosis, the normal chromosome complement is 2n or diploid. In humans, the diploid chromosome number is 46 while in *Drosophila* it is 8. *Essentials*-Tab.2.1 lists the *haploid* chromosome number for a variety of species. Notice that in man and flies, the haploid chromosome number is one-half the diploid number. This applies to other organisms as well. However, it is very important to realize that *haploidy* specifically refers to the fact that each haploid cell contains *one chromosome of each homologous pair of chromosomes.* Compare the nuclear contents of a spermatid and a cell at zygonema in the *Essentials* text. Note that each spermatid contains one member of each of the original chromosome pairs (seen at zygonema). Haploidy of usually symbolized as *n*. The change from a diploid (2n) to haploid (n) occurs during *reduction division* when tetrads become dyads during meiosis I. Referring to the number of human chromosomes, the primary spermatocyte (2n=46) becomes two secondary spermatocytes each with n = 23.

3. As you examine the criteria for *homology* in question #2 above, you can see that overall length and centromere position are but two factors required for homology. Most importantly, genetic content in non-homologous chromosomes is expected to be quite different. Other factors including banding pattern and time of replication during S phase would also be expected to vary among non-homologous chromosomes.

4. Because much of Chapter #2 deals with meiosis, it would be best to deal with this question by reading the sections which cover meiosis in the *Essentials* text. Understanding meiosis and all the related terms is essential for an understanding of genetics. There are several sample test questions which will help you determine your understanding of meiosis.

5. The first sentence tells you that 2n = 16 and it is a question about mitosis. Since each chromosome in prophase is doubled (having gone through an S phase) and is visible at the end of prophase, there should be 32 chromatids. Because the centromeres divide and what were previously sister chromatids migrate to opposite poles during anaphase, there should be 16 chromosomes moving to each pole. If you refer to F2.1 and the *Essentials* text you will see an example which will help illustrate these points.

6. Refer to the *Essentials* text (Fig. 2.3) for an explanation. Notice the different anaphase shapes in the figures. Your understanding of these structures will be determined by several of the sample test questions at the end of this section.

7. Because of a cell wall around the plasma membrane in plants, a cell plate, which was laid down during anaphase, becomes the middle lamella where primary and secondary layers of the cell wall are deposited.

8. Carefully read the section dealing with mitosis in *Essentials*. Refer to F2.1 and figures in the *Essentials* text for information pertaining to the interphase. Refer to the *Essentials* text (Figs. 7A, 7B) for a diagram of mitosis. Notice that, in contrast to meiosis, there is no pairing of homologous chromosomes in mitosis and the chromosome number does not change (see T2.1).

9. Not necessarily. If crossing over occurred in meiosis I, then the chromatids in the secondary oocyte are not identical. Once they separate during meiosis II, unlike chromatids reside in the ootid and the second polar body.

10. Compared with mitosis, meiosis provides for a reduction in chromosome number, and an opportunity for exchange of genetic material from homologous chromosomes. In mitosis there is no change in chromosome number (see T2.1) or kind in the two daughter cells whereas in meiosis numerous potentially different haploid (n) cells are produced. During oogenesis, only one of the four meiotic products is functional; however, four of the four meiotic products of spermatogenesis are potentially functional.

11. (a) *Synapsis* is the point-by-point pairing of homologous chromosomes during prophase of meiosis I.

(b) *Bivalents* are those structures formed by the synapsis of homologous chromosomes. In other words, there are two chromosomes (four chromatids) which make up a bivalent. If an organism has a diploid chromosome number of 46, then there will be 23 bivalents in meiosis I.

(c) *Chiasmata* is the plural form of chiasma and refers to the structure, when viewed microscopically, of crossed chromatids. Notice in the figures of the *Essentials* text the exchange of chromatid pieces in diplonema and diakinesis.

(d) *Crossing over* is the exchange of genetic material between chromatids. Also called recombination, it is a method of providing genetic variation through the breaking and rejoining of chromatids. Notice in *Essentials* the mixing of genetic material along the length of the chromatids.

(e) *Chromomeres* are patches of chromatin which look different from neighboring patches along the length of a chromosome.

(f) Examine F2.1 in this book. Notice that *sister chromatids* are "post-S phase" structures of replicated chromosomes. Sister chromatids are genetically identical (except where mutations have occurred) and are attached to the same centromere. Identify the sister chromatids in figures in the *Essentials* text. Note that sister chromatids separate from each other during anaphase of mitosis and anaphase II of meiosis.

(g) Tetrads are synapsed homologous chromosomes thereby composed of four chromatids. There are as many tetrads as the haploid chromosome number.

(h) Actually, each tetrad is made of two dyads which separate from each other during anaphase I of meiosis. Dyads are composed of two chromatids joined by a centromere.

(i) At anaphase II of meiosis, the centromeres divide and sister chromatids go to opposite poles.

(j) The *synaptonemal complex* is a nuclear structure which is often found associated with synapsed meiotic chromosomes. It looks and acts like a zipper.

12. Look carefully at F2.3 in this book and notice that, for a cell with 4 chromosomes, there are two tetrads each comprised of a homologous pair of chromosomes.

(a) If there are 16 chromosomes there should be 8 tetrads.

(b) Also note that, after meiosis I and in the second meiotic prophase there are as many dyads as there are *pairs* of chromosomes. There will be 8 dyads.

(c) Because the monads migrate to opposite poles during meiosis II (from the separation of dyads) there should be 8 monads migrating to *each* pole.

13. Examine the *Essentials* text and the figure on the next page. Notice that major differences include the sex in which each occurs, and that the distribution of cytoplasm is unequal in oogenesis but considered to be equal in the products of spermatogenesis. Chromosomal behavior is the same in spermatogenesis and oogenesis except that the nuclear activity in oogenesis is "off-center" thereby producing first and second polar bodies by unequal cytoplasmic division. Each spermatogonium and primary spermatocyte produces four spermatids whereas each oogonium and primary oocyte produces one ootid. Because early development occurs in the absence of outside nutrients, it is likely that the unequal distribution of cytoplasm in oogenesis evolved to provide sufficient information and nutrients to support development until the transcriptional activities of the zygotic nucleus begin to provide products.

Polar bodies probably represent non-functional by-products of such evolution.

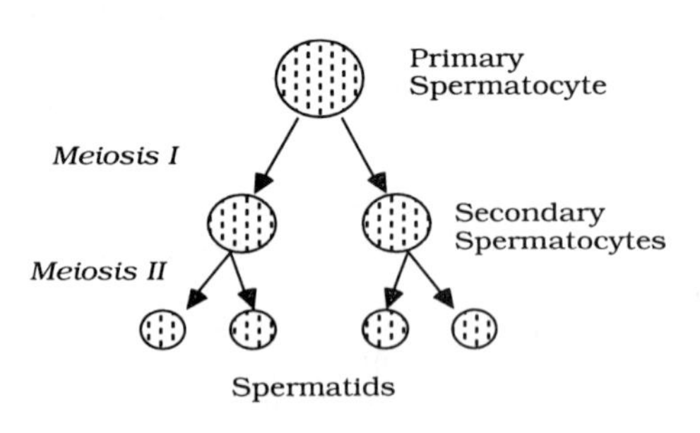

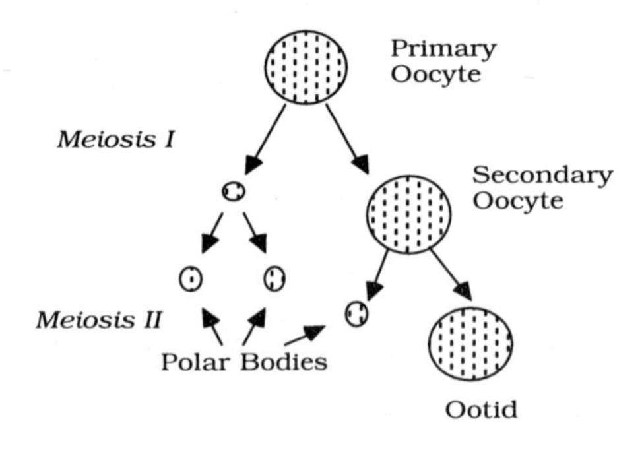

14. In meiosis, various chromosomal arrangements are possible because of random alignment of homologous chromosomes at metaphase I. In addition, crossing over, which introduces additional variation, is virtually absent in mitotic processes.

15. This question specifically tests your understanding of meiosis and the behavior of chromosomes during anaphase. In this question you must first visualize the alignment of the three homologous chromosome pairs C1,C2, M1,M2, and S1,S2 in mitosis (no synapsis of homologues) as compared with the alignment in meiosis.

(a)

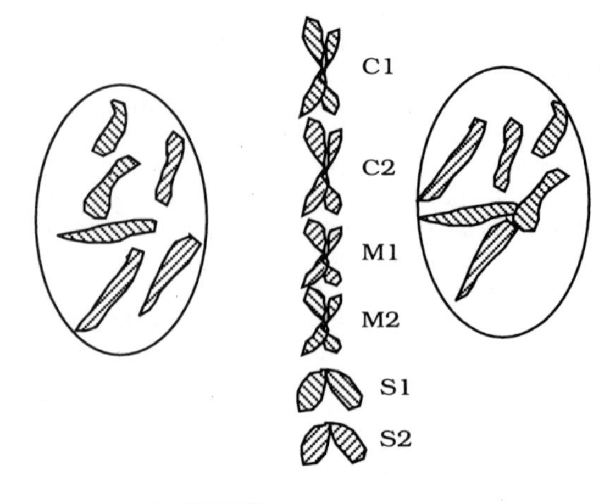

Paternally derived

Maternally derived

The two daughter cells will have the same chromosomes after mitosis

(b)

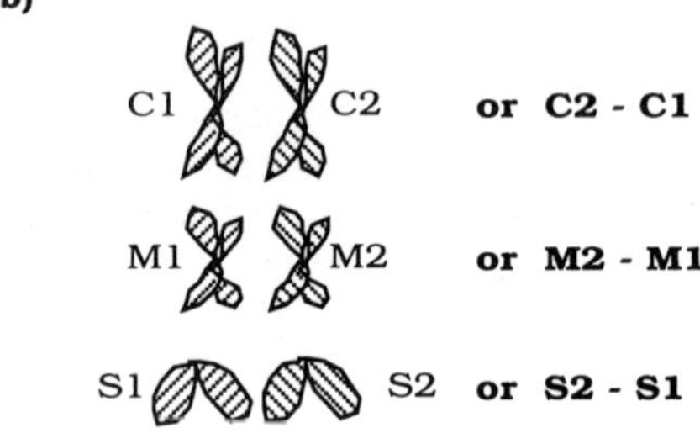

Paternally derived

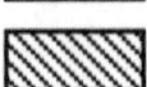
Maternally derived

Notice that there are no constraints on the alignment of different pairs of homologous chromosomes, therefore one could list 8 configurations.

c) Because of the independent assortment of non-homologous chromosomes at anaphase I and the fact that anaphase II separates sister (identical) chromatids, there will be eight different meiotic (haploid) products.

16. There would be 16 combinations possible.

17. In plants, one often speaks of "alternation of generations" in which the life cycle alternates between the diploid sporophyte and the haploid gametophyte stages. Meiosis and fertilization bridge the sporophyte and gametophyte generations.

18. The cytological origin of the mitotic chromosome (and the meiotic chromosome for that matter) appears during the prophase through a condensation of the dispersed chromatin fibers present during interphase. During interphase, chromatin consists of invisible threads (under light microscopy) of DNA associated with histones. In this form, DNA can function in transcription and replication. Visible mitotic chromosomes appear when each chromosome coils and condenses according to the current *folded fiber model.*

19. *Polytene chromosomes* are specialized chromosomes which occur in a variety of organisms, namely, dipteran larval cells (salivary, midgut, rectal, and malpighian). Visible by relatively low magnification light microscopy and having pronounced bands (chromomeres), they have been of great benefit to cytogeneticists. Polytene chromosomes are observed at interphase when transcription typically occurs, thus allowing scientists to observe active genes (puffs) under certain circumstances.

Lampbrush chromosomes are meiotic, prophase I chromosomes which exist in most vertebrate oocytes and spermatocytes of some insects. In the vertebrate oocyte, such chromosomes are active in metabolic activities of the cell. Lateral loops extend from central axes of lampbrush chromosomes and RNA is actively synthesized from such loops. In some ways the loops may be thought of as similar to the puffs of polytene chromosomes. Both forms of chromosomes (polytene and lampbrush) have provided significant information as to chromosome structure and certain developmental processes.

20. *cdc*2 is the symbol given to a cell division cycle gene which produces a kinase that couples with various cyclins to direct cells passed certain critical points (G1/S and G2/M) of the cell cycle. p53 represents a 53 kilodalton protein which acts as a tumor suppressor by regulating the transition between G1 and S.

21. It is through mutations in cell cycle control genes that scientists are allowed to determine how normal genes function. For example, when mutations cause cells to stop at the beginning of mitosis, it can be inferred that certain gene products are required for passage into mitosis.

3

Mendelian Genetics

Vocabulary: Organization and Listing of Terms

Historical

Mendelian Genetics (Gregor Mendel) - 1866

- *Pisum sativum*
  - units of inheritance (particulate)
- Rebirth (1900)
  - Hugo DeVries
  - Karl Correns
  - Erich Tschermak
- Walter Sutton
- Theodor Boveri

Structures and Substances

Unit factors

Genes

Alleles

Processes/Methods

Transmission genetics

- monohybrid cross
  - maternal parent
  - paternal parent

Punnett squares

- test cross
- parental generation (P_1)
- first filial generation (F_1)
- second filial generation (F_2)
- ratios
  - 3:1
  - 9:3:3:1
  - 27:9:9:9:3:3:3:1
- product law
  - 2^n (n = haploid chromosome number)
- dihybrid cross (two-factor cross)
- trihybrid cross (three-factor cross)

- forked-line (branch diagram) method
- Statistical testing (analysis)
  - predicted occurrences
  - proportions
  - sample size
  - chance deviation
  - random fluctuations
  - null hypothesis
  - measured values
  - predicted values
  - goodness of fit
  - chi-square analysis (χ^2)
  - degrees of freedom
  - probability value (p)
    - reject the null hypothesis
    - fail to reject the null hypothesis
    - 0.05 probability value
- Pedigree
  - sibs, sibship line
  - monozygotic (identical) twins
  - dizygotic (fraternal) twins
  - proband (propositus)

Concepts

- Mendel's Postulates
  - unit factors in pairs (F3.1)
  - dominance/recessiveness (F3.1)
  - segregation
  - independent assortment
  - symbolism (F3.2)
  - phenotype
  - genotype
  - homozygous (homozygote)
  - heterozygous (heterozygote)
- Chromosome theory of inheritance
- Homologous chromosomes

F 3.1. Illustration of the union of maternal and paternal genes (*A* and *a*) to give two genes in the zygote. Mendelian "unit factors" occur in pairs in diploid organisms. Dominant genes are often given the upper case letter as the symbol while the lower case letter is often used to symbolize the recessive gene.

It is important to see that each parent contributes one chromosome of each type (homologue) and thus one gene of each gene pair.

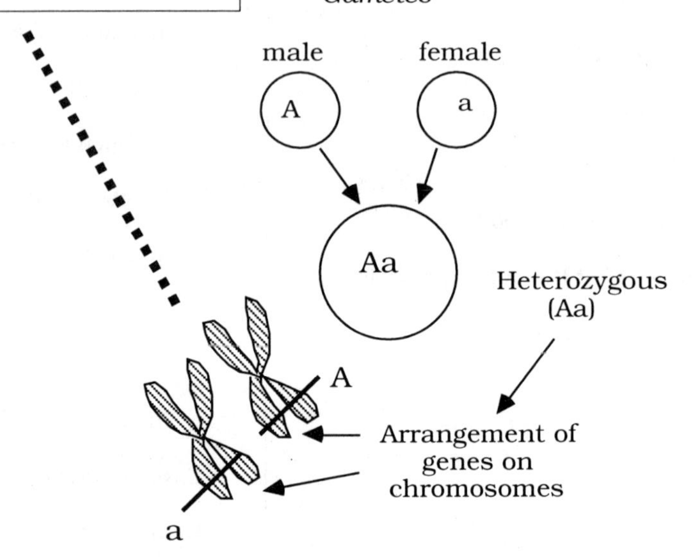

F3.2. It is important to see that in the drawing below there are four variations, but only two characteristics (shape and color). When developing symbols for gene pairs, keep upper and lower case letters (or other compatible scheme) *of the same letter* to represent genes of the same characteristic.

Two characteristics but a total of four alternatives

Two characteristics

1. Shape (round, wrinkled)
2. Color (black, white)

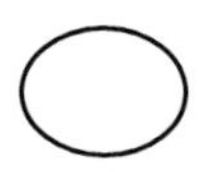

Symbol choices (examples)

W = round w = wrinkled
(assuming round is dominant to wrinkled)

B = black b = white
(assuming black is dominant to white)

Selection of appropriate symbolism is critical

Solutions to Problems and Discussion Questions

1. Several points surface in the first sentence of this question. First, two alternatives (black and white) of one characteristic (coat color) are being described, therefore a monohybrid condition (Fig.3.1).

Second, are the guinea pigs in the parental generation (P_1) homozygous or heterozygous? Notice in the introductory sentence, just after PROBLEMS AND DISCUSSION QUESTIONS, there is the statement "members of the P_1 generation are homozygous..."

Third, which is dominant, *black* or *white*? Note that all the offspring are black, therefore black can be considered dominant. The second sentence of the problem verifies that a monohybrid cross is involved because of the 3/4 black and 1/4 white distribution in the offspring. Knowing that genes occur in pairs in diploid organisms, one can write the genotypes and the phenotypes as follows:

P_1:

Phenotypes: Black X White

Genotypes: *WW* *ww*

Gametes: *W* *w*

F_1: *Ww* (Black)

F_1 X F_1:

Phenotypes: Black X Black

Genotypes: *Ww* *Ww*

Gametes: *W* *w* *W* *w*

(combine as in *Essentials*)

F_2:

Phenotypes: Black Black Black White

Genotypes: *WW* *Ww* *Ww* *ww*

2. Start out with the following gene symbols;

A =normal (not albino),

a =albino.

Since albinism is inherited as a recessive trait, genotypes *AA* and *Aa* should produce the normal phenotype, while *aa* will give albinism.

(a) The parents are both normal, therefore they could be either *AA* or *Aa*. The fact that they produce an albino child requires that each parent provides an *a* gene to the albino child; thus the parents must both be heterozygous(*Aa*).

(b) To start out, the normal male could have either the *AA* or *Aa* genotype. The female must be *aa*. Since all the children are normal one would consider the male to be *AA* instead of *Aa* . However, the male *could* be *Aa*. Under that circumstance, the likelihood of having six children, all normal is 1/64.

3. *Unit factors in Pairs, Dominance and Recessiveness, Segregation*

4. *Pisum sativum* is easy to cultivate. It is naturally self-fertilizing, but it can be crossbred. It has several visible features (*e.g.*, tall or short, red flowers or white flowers) which are consistent under a variety of environmental conditions yet contrast due to genetic circumstances. Seeds could be obtained from local merchants.

5. With many problems, students often have trouble getting started in the right direction and seeing the problem through to the necessary conclusions. First, read the entire question and see that you are to determine (1) the pattern of inheritance for "checkered and plain," and (2) the gene symbols and genotypes of all the parents and offspring. Notice that there is reference to one characteristic, *pattern*, with two alternatives, checkered vs. plain. We should consider this to be a monohybrid condition unless complications arise.

Assignment of symbols:

P = checkered; *p* = plain. Checkered is tentatively assigned the dominant function because in a casual examination of the data, especially cross (b), we see that checkered types are more likely to be produced than plain types.

Cross (a):

PP X *PP* or *PP* X *Pp*

Cross (b):

PP X *pp*

This assignment seems reasonable because among 38 offspring, no plain types are produced. In addition, we would expect all the F_1 progeny to be heterozygous.

Cross (c):

Because all the offspring from this cross are plain, there is no doubt that the genotype of both parents is *pp*.

Genotypes of all individuals:

P_1 Cross	F_1 Progeny: Checkered	F_1 Progeny: Plain
(a) *PP* X *PP*	*PP*	
(b) *PP* X *pp*	*Pp*	
(c) *pp* X *pp*		*pp*

In a mating of the F1 X F1 from cross (b), one would expect a 3:1 ratio of checkered to plain as shown below:

	P	*p*
P	*PP*	*Pp*
p	*Pp*	*pp*

6. In the first sentence you are told that there are two *characteristics* which are being studied; seed shape and cotyledon color. Expect, therefore, this to be a dihybrid situation with *two gene pairs* involved. One also sees the possible alternatives of these two characteristics: *seed shape*; wrinkled vs. round; *cotyledon color*; green vs. yellow. After reading the second sentence you can predict that the gene for round seeds is dominant to that for wrinkled seeds and the gene for yellow cotyledons is dominant to the gene for green cotyledons.

Symbolism:

w = wrinkled seeds   *g* = green cotyledons

W = round seeds   *G* = yellow cotyledons

P_1:

WWGG X *wwgg*

Parents are considered to be homozygous for two reasons. First, in the introductory sentence, just after PROBLEMS AND DISCUSSION QUESTIONS, there is the statement "members of the P_1 generation are homozygous..." Second, notice that the only offspring are those with round seeds and yellow cotyledons.

Gametes produced: One member of each gene pair is "segregated" to each gamete.

WWGG   *wwgg*

(*WG*)   (*wg*)

F_1: *WwGg*

F_1 X F_1:

WwGg X *WwGg*

Gametes produced: Under conditions of independent assortment, there will be four (2^n, where n = number of heterozygous gene pairs) different types of gametes produced by each parent.

Punnett Square (*Essentials*-Fig.3.7):

	WG	*Wg*	*wG*	*wg*
WG	*WWGG*	*WWGg*	*WwGG*	*WwGg*
Wg	*WWGg*	*WWgg*	*WwGg*	*Wwgg*
wG	*WwGG*	*WwGg*	*wwGG*	*wwGg*
wg	*WwGg*	*Wwgg*	*wwGg*	*wwgg*

Collecting the phenotypes according to the dominance scheme presented above, gives the following:

9/16	*W_G_*	round seeds, yellow cotyledons
3/16	*W_gg*	round seeds, green cotyledons
3/16	*wwG_*	wrinkled seeds, yellow cotyledons
1/16	*wwgg*	wrinkled seeds, green cotyledons

Notice that a dash (_) is used where, because of dominance, it makes no difference as to the dominance/recessive status of the allele.

Forked, or branch diagram:

Seed shape	*Cotyledon color*	*Phenotypes*
3/4 round	3/4 yellow	9/16 round,yellow
	1/4 green	3/16 round, green
1/4 wrinkled	3/4 yellow	3/16 wrinkled, yellow
	1/4 green	1/16 wrinkled, green

7. Symbolism as before:

w = wrinkled seeds	*g* = green cotyledons
W = round seeds	*G* = yellow cotyledons

Examine each characteristic (seed shape vs. cotyledon color) separately.

(a) Notice a 3:1 ratio for seed shape, therefore *Ww* X *Ww*; and no green cotyledons, therefore *GG* X *GG* or *GG* X *Gg*. Putting the two characteristics together gives

WwGG X *WwGG*

or

WwGG X *WwGg*

(b) Notice a 1:1 ratio for seed shape (8/16 wrinkled and 8/16 round) and a 3:1 ratio for cotyledon color (12/16 yellow and 4/16 green). Therefore the answer is

wwGg X *WwGg*.

(c) This is a typical 1:1:1:1 test cross (or backcross) ratio:

WwGg X *wwgg*

8. A test cross involves a fully heterozygous organism mated with a fully homozygous recessive organism. In Problem #7, (c) is a test cross.

9. Because *independent assortment* may be defined as one gene pair segregating independently of another gene pair, one would need at least two gene pairs in order to demonstrate independent assortment, as is the case in Problem #7.

10. Mendel's four postulates are related to the diagram below.

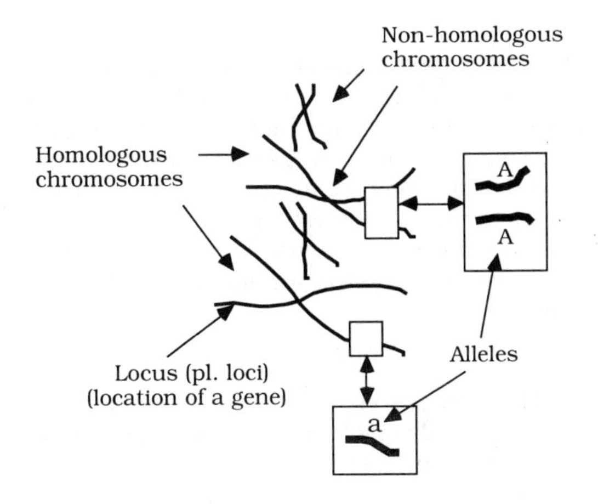

1. Factors occur in pairs. Notice *A* and *a*.

2. Some genes are dominant to their alleles. Notice *A* and *a*.

3. Alleles segregate from each other during gamete formation. When homologous chromosomes separate from each other at anaphase I, alleles will go to opposite poles of the meiotic apparatus.

4. One gene pair separates independently from other gene pairs. Different gene pairs on the same homologous pair of chromosomes (if far apart) or on non-homologous chromosomes will separate independently from each other during meiosis.

11. Carefully re-read the answer to Question #2 in Chapter #2. Briefly, the factors which specify chromosomal homology are the following:

overall length

position of the centromere

banding patterns

type and location of genes

autoradiographic pattern

12. Homozygosity refers to a condition where both genes of a pair are the same (i.e. *AA* or *GG* or *hh*), whereas heterozygosity refers to the condition where members of a gene pair are different (i.e. *Aa* or *Gg* or *Bb*). Homozygotes produce only one type of gamete whereas heterozygotes will produce 2^n types of gametes where n = number of heterozygous gene pairs (assuming independent assortment).

13. There are two characteristics presented here, body color and wing length. First, assign meaningful gene symbols.

Body color	*Wing length*
E = grey body color	*V* = long wings
e = ebony body color	*v* = vestigial wings

(a)

P_1:

EEVV X *eevv*

F_1: *EeVv* (grey, long)

F_2:

This will be the result of a Punnett Square with 16 boxes as in *Essentials*-Fig.3.7.

Phenotypes	*Ratio*	*Genotypes*	*Ratio*
grey , long	9/16	*EEVV*	1/16
		EEVv	2/16
		EeVV	2/16
		EeVv	4/16
grey, vestigial	3/16	*EEvv*	1/16
		Eevv	2/16
ebony, long	3/16	*eeVV*	1/16
		eeVv	2/16
ebony, vestigial	1/16	*eevv*	1/16

(b)

P_1:

EEvv X *eeVV*

F_1: It is important to see that the results from this cross will be exactly the same as those in part (a) above. The only difference is that the recessive genes are coming from both parents, rather than from one parent only as in (a). The F_2 ratio will be the same as (a) also. When you have genes on the autosomes (not X-linked), independent assortment, complete dominance, and no gene interaction (see later) in a cross involving double heterozygotes, the offspring ratio will be in the ratio 9:3:3:1.

(c)

P_1:

EEVV X EEvv

F_1: *EEVv* (grey, long)

F_2: Notice that all the offspring will have grey bodies and you will get a 3:1 ratio of long to vestigial wings. You should see this before you even begin working through the problem. Even though this cross involves two gene pairs, it will give a "monohybrid " type of ratio because one of the gene pairs is homozygous (body color) and **one** gene pair is heterozygous (wing length).

Phenotypes	*Ratio*	*Genotypes*	*Ratio*
grey, long	3/4	*EEVV*	1/4
		EEVv	2/4
grey, vestigial	1/4	*EEvv*	1/4

NOTE: After working through this problem, it is important that you try to work similar problems without constructing the time-consuming Punnett squares, especially if each problem asks for phenotypic rather than genotypic ratios.

14. The general formula for determining the number of kinds of gametes produced by an organism is 2^n where n = number of *heterozygous* gene pairs.

(a) 4: *AB, Ab, aB, ab*

(b) 2: *AB, aB*

(c) 8: *ABC, ABc, AbC, Abc, aBC, aBc, abC, abc*

(d) 2: ABc, aBc

(e) 4: ABc, Abc, aBc, abc

(f) 2^5= 32

ABCDE	aBCDE
ABCDe	aBCDe
ABCdE	aBCdE
ABCde	aBCde
ABcDE	aBcDE
ABcDe	aBcDe
ABcdE	aBcdE
ABcde	aBcde
AbCDE	abCDE
AbCDe	abCDe
AbCdE	abCdE
AbCde	abCde
AbcDE	abcDE
AbcDe	abcDe
AbcdE	abcdE
Abcde	abcde

Notice that there is a pattern that can be used to write these gametes so that fewer errors will occur.

15.

(a) When examining this cross

AaBbCc X AaBBCC

expect there to be eight different kinds of gametes from one parent (*AaBbCc*), and two different kinds from the other (*AaBBCC*). Therefore there should be sixteen kinds (genotypes) of offpsring (8 X 2).

Gametes:	Gametes:
ABC	*ABC*
ABc	*aBC*
AbC	
Abc	
aBC	
aBc	
abC	
abc	

Offspring:

Genotypes	*Ratio*	*Phenotypes*
AABBCC	(1/16)	
AABBCc	(1/16)	
AABbCC	(1/16)	
AABbCc	(1/16)	
AaBBCC	(2/16)	A_B_C_ = 12/16
AaBBCc	(2/16)	
AaBbCC	(2/16)	
AaBbCc	(2/16)	
aaBBCC	(1/16)	
aaBBCc	(1/16)	aaB_C_ = 4/16
aaBbCC	(1/16)	
aaBbCc	(1/16)	

(b) There will be four kinds of gametes for the first parent (AaBBCc) and two kinds of gametes for the second parent.

Gametes: Gametes:

Gametes	Gametes
ABC	*aBC*
ABc	*aBc*
aBC	
aBc	

Offspring:

Genotypes	*Ratio*	*Phenotypes*	
AaBBCC	1/8	*A_BBC_*	= 3/8
AaBBCc	2/8		
AaBBcc	1/8	*A_BBcc*	= 1/8
aaBBCC	1/8	*aaBBC_*	= 3/8
aaBBCc	2/8		
aaBBcc	1/8	*aaBBcc*	= 1/8

(c) There will be eight (2^n) different kinds of gametes from each of the parents, therefore a 64-box Punnett square. Doing this problem by the forked-line method helps considerably.

```
                         / 1/4 CC = 1/64 AABBCC
              1/4 BB  <  — 2/4 Cc = 2/64 AABBCc
             /           \ 1/4 cc
            /            / 1/4 CC          etc.
1/4 AA  —  2/4 Bb  <  — 2/4 Cc
            \            \ 1/4 cc
             \           / 1/4 CC
              1/4 bb  <  — 2/4 Cc
                         \ 1/4 cc

                       1/4 CC
            1/4 BB     2/4 Cc
                       1/4 cc
                       1/4 CC
2/4 Aa      2/4 Bb     2/4 Cc
                       1/4 cc
                       1/4 CC
            1/4 bb     2/4 Cc
                       1/4 cc

                       1/4 CC
            1/4 BB     2/4 Cc
                       1/4 cc
                       1/4 CC
1/4 aa      2/4 Bb     2/4 Cc
                       1/4 cc
                       1/4 CC
            1/4 bb     2/4 Cc
                       1/4 cc
```

Simply multiply through each component to arrive at the final genotypic frequencies.

For the phenotypic frequencies, set up the problem in the following manner.

```
                         /3/4 C_      = 27/64 A_B_C_
              3/4 B_  —1/4 cc         = 9/64  A_B_cc
             /
3/4 A_    — 1/4 bb — 3/4 C_           etc.
                     \ 1/4 cc

1/4 aa    — 3/4 B_ — 3/4 C_
                     \ 1/4 cc
          \
            1/4 bb — 3/4 C_
                     \ 1/4 cc
```

16. In reading this question, notice that there are two characteristics being considered; seed color (yellow, green) and seed shape (round, wrinkled). At this point you should be able to do this problem without writing down each of the steps. The F_1 can be considered to be a double heterozygote (with round and yellow being dominant). See the cross this way:

Symbols:

Seed shape	*Seed color*
W = round	*G* = yellow
w = wrinkled	*g* = green

P_1:

WWgg X *wwGG*

F_1: *WwGg* cross to *wwgg*

(which is a typical test cross)

The offspring will occur in a typical 1:1:1:1 as

1/4 *WwGg* (round, yellow)

1/4 *Wwgg* (round, green)

1/4 *wwGg* (wrinkled, yellow)

1/4 *wwgg* (wrinkled, green)

Again, at this point it would be very helpful if you could do such simple problems by inspection.

17. Since these are F_2 results from monohybrid crosses, a 3:1 ratio is expected for each. Referring to the *Essentials* text one can set up the analysis easily.

(a)

Expected ratio	*Observed (o)*	*Expected (e)*
3/4	882	885.75
1/4	299	295.25

Expected values are derived by multiplying the expected ratio by the total number of organisms.

$$\chi^2 = \Sigma \frac{(o - e)2}{e} = 0.064$$

In looking in the χ^2 table in the *Essentials* text, with 1 degree of freedom (because there were two classes, therefore n-1 or 1 degree of freedom), we find a probability (*P*) value between 0.9 and 0.5.

We would therefore say that there is a "good fit" between the observed and expected values. Notice that as the deviations between the observed and expected values increase, the value of χ^2 increases. So the higher the χ^2 value the more likely the null hypothesis will be rejected.

(b)

Expected ratio	*Observed (o)*	*Expected (e)*
3/4	705	696.75
1/4	224	232.25

$\chi^2 = 0.39$

The *P* value in the table for 1 degree of freedom is still between 0.9 and 0.5, however because the χ^2 value is larger in (b) we should say that the deviations from expectation are greater.

18. One must think of this problem as a dihybrid F_2 situation with the following expectations:

Expected ratio	*Observed (o)*	*Expected (e)*
9/16	315	312.75
3/16	108	104.25
3/16	101	104.25
1/16	32	34.75

$\chi^2 = 0.5$

Looking at the table in *Essentials* one can see that this χ^2 value is associated with a probability greater than 0.90 for 3 degrees of freedom (because there are now four classes in the χ^2 test). The observed and expected values do not deviate significantly.

To deal with parts **(b)** and **(c)** it is easier to see the observed values for the monohybrid ratios if the phenotypes are listed:

smooth, yellow	315
smooth, green	108
wrinkled, yellow	101
wrinkled, green	32

For the smooth: wrinkled *monohybrid component*, the smooth types total 423 (315 + 108), while the wrinkled types total 133 (101 + 32).

Expected ratio	*Observed (o)*	*Expected (e)*
3/4	423	417
1/4	133	139

The χ^2 value is 0.35 and in examining the table in *Essentials* for 1 degree of freedom, the P value is greater than 0.50 and less than 0.90. We fail to reject the null hypothesis and are confident that the observed values do not differ significantly from the expected values.

(c) For the yellow:green portion of the problem, see that there are 416 yellow plants (315 + 101) and 140 (108 + 32) green plants.

Expected ratio	*Observed (o)*	*Expected (e)*
3/4	416	417
1/4	140	139

The χ^2 value is 0.01 and in examining the table in *Essentials* for 1 degree of freedom, the P value is greater than 0.90. We fail to reject the null hypothesis and are confident that the observed values do not differ significantly from the expected values.

19. It would be best to set up two tables based on the two hypotheses:

Expected ratio	*Observed (o)*	*Expected (e)*
3/4	250	300
1/4	150	100

Expected ratio	*Observed (o)*	*Expected (e)*
1/2	250	200
1/2	150	200

For the test of a 3:1 ratio, the χ^2 value is 33.3 with an associated P value of less than 0.01 for 1 degree of freedom.

For the test of a 1:1 ratio, the χ^2 value is 25.0 again with an associated P value of less than 0.01 for 1 degree of freedom. Based on these probability values, both null hypotheses should be rejected.

20. Use of the P = 0.10 as the "critical" value for rejecting or failing to reject the null hypothesis instead of P = 0.05 would allow more null hypotheses to be rejected. Notice in the *Essentials* text that as the χ^2 values increase, there is a higher likelihood that the null hypothesis will be rejected because the higher values are more likely to be associated with a P value which is less than 0.05.

As the critical P value is increased, it takes a smaller χ^2 value to cause rejection of the null hypothesis. It would take less difference between the expected and observed values to reject the null hypothesis, therefore the stringency of failing to reject the null hypothesis is increased.

21. While there are many different inheritance patterns which will be described later in the *Essentials* text (codominance, incomplete dominance, sex-linked inheritance, etc.), the range of solutions to this question is limited to the concepts developed in the first three chapters, namely dominance or recessiveness.

If a gene is dominant, it will not skip generations nor will it be passed to offspring unless the parents possess the gene. On the other hand, genes which are recessive can skip generations and exist in a carrier state in parents. For example, notice that II-4 and II-5 produce a female child (III-4) with the affected phenotype. On these criteria alone, the gene must be viewed as being recessive. Note: if a gene is recessive and X-linked (to be discussed later) the pattern will often be from affected male to carrier female to affected male.

To provide genotypes for each individual consider that if the box or circle is shaded, the *aa* genotype is to be assigned. If offspring are affected (shaded) a recessive gene must have come from both parents.

I-1 (*aa*), I-2 (*Aa*), I-3 (*Aa*), I-4(*Aa*)

II-1 (*aa*), II-2 (*Aa*), II-3 (*aa*), II-4 (*Aa*), II-5 (*Aa*), II-6 (*aa*), II-7 (*AA* or *Aa*), II-8 (*AA* or *Aa*)

III-1 (*AA* or *Aa*), III-2 (*AA* or *Aa*), III-3 (*AA* or *Aa*), III-4 (*aa*), III-5 (probably *AA*), III-6 (*aa*)

IV-1 through IV-7 all *Aa*.

22. *Unit Factors in Pairs*: It is important to see that each time a phenotype (normal or abnormal) is being stated, genotypes are symbolized as pairs of genes; *AA*, *Aa* or *aa*. Review F3.2 to understand the need to assign appropriate symbols to genes.

Dominance and Recessiveness: Because the gene for non-shaded is completely dominant over the gene for shaded (say *a* is fully recessive), it was necessary to consider, at first, whether non-shaded individuals in the problem were homozygous normal (*AA*) or heterozygous (*Aa*). By looking at the frequency of expression of the recessive gene in the offspring (in *aa* individuals), one can often determine an *Aa* type from an *AA* type.

Segregation: During gamete formation when homologous chromosomes move to opposite poles, paired elements (genes) separate from each other.

23. Appling the same logic as in a previous pedigree question, the gene is inherited as an autosomal recessive. Notice that two normal individuals in II-3 and II-4 have produced a daughter (III-2) with myopia.

I-1 (*AA* or *Aa*), I-2 (*aa*), I-3 (*Aa*), I-4 (*Aa*)

II-1 (*Aa*), II-2 (*Aa*), II-3 (*Aa*), II-4(*Aa*), II-5 (*aa*), II-6 (AA or Aa), II-7 (AA or Aa)

III-1 (*AA* or *Aa*), III-2 (*aa*), III-3 (*AA* or *Aa*)

24. (a) First consider that each parent is homozygous (true-breeding in the question) and since in the F_1 only round, axial, violet, and full phenotypes were expressed, they must each be dominant.

(b) Round, axial, violet and full would be the most frequent phenotypes:

3/4 X 3/4 X 3/4 X 3/4

(c) Wrinkled, terminal, white, and constricted would be the least frequent phenotypes:

1/4 X 1/4 X 1/4 X 1/4

(d) 3/4 X 1/4 X 3/4 X 1/4

(e) There would bc eight different phenotypes in the test cross offspring just as there are eight different phenotypes in the F_2 generation.

25. (a) The first task is to draw out an accurate pedigree (one of serveral possibilities):

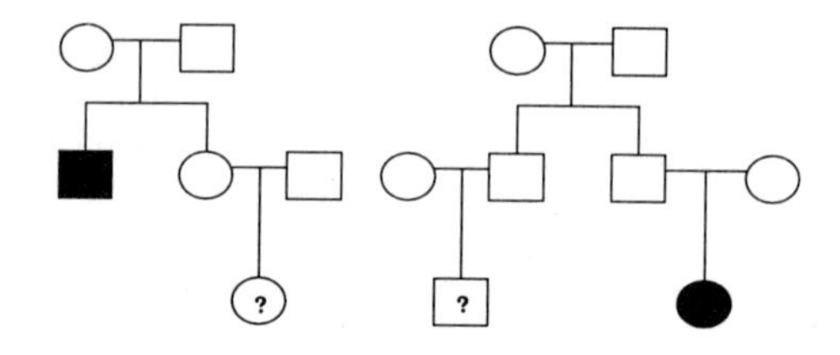

(b) The probability that the female (whose maternal uncle had TSD) is heterozygous is 1/3 because she is not TSD and her mother had a 2/3 chance of being heterozygous. Given that her mother is heterozygous, she has a 1/2 chance of passing the TSD gene to her daughter (2/3 X 1/2 = 1/3). The male (whose paternal first cousin had TSD) has a 1/4 chance of being heterozygous, assuming that either (but not both, because the gene is said to be rare) his grandmother or grandfather was heterozygous. Therefore the probability that both the male and female are heterozygous is:

1/3 X 1/4 = 1/12

(c) The probability that neither is heterozygous is:

2/3 X 3/4 = 6/12

(d) The probability that one is heterozygous is:

(1/3 X 3/4) + (2/3 X 1/4) = 5/12

4

Modification of Mendelian Ratios

Vocabulary: Organization and Listing of Terms

Structures and Substances

Allele

Native antigens

- glycoproteins

X chromosome

Y chromosome

Processes/Methods

Incomplete (partial) dominance

- pink flowers
- 1:2:1 phenotypic ratio

Codominance

- $L^M L^N$ alleles (MN blood groups)
- 3:6:3:1:2:1

Multiple alleles

- I^A I^B I^O (ABO blood groups)
- antigen-antibody reaction
  - isoagglutinogen
  - H substance
  - Bombay phenotype
- *white eye* in *Drosophila*

Lethal alleles

- recessive
- dominant
  - yellow coat color in mice
  - Huntington disease

Gene interaction: discontinuous variation

- epistasis
- hypostatic
  - homozygous recessive
    - coat color in mammals
    - Bombay phenotype
    - 9:3:4

dominant

fruit color in squash

12:3:1

other

white flowers in peas

9:7

novel phenotypes

fruit shape in *Cucurbita*

Gene interaction: continuous variation

Sex linkage

hemizygous

crisscross pattern

Sex-limited inheritance

Sex-influenced inheritance

Concepts

Gene interaction (F4.1)

Neo-Mendelian genetics

Allele (F4.2)

wild type

mutation

loss of wild type function

reduced or increased function

Symbolism (F4.2)

recessive trait (D, d; e, e^+)

dominant trait (Wr, Wr^+)

"+" as a superscript

+ as a symbol for wild type (+/*e*)

no dominance (R^1, R^2)

Modified ratios (T4.1)

3:1

1:2:1

9:3:3:1

3:6:3:1:2:1

9:3:4

12:3:1

9:7

1:4:6:4:1

Other patterns of inheritance

Sex linkage

Sex-limited

Sex-influenced

T4.1. Examples of typical monohybrid and dihybrid ratios and several modifications.

Basic Ratio	*Modification*	*Explanation*
3:1	1:2:1	Incomplete dominance
		Codominance
9:3:3:1	9:3:4	Epistasis
	12:3:1	Epistasis
	9:7	Epistasis
	3:6:3:1:2:1	Dominance + incomplete dominance or codominance

F4.1. Illustration of gene interaction where products from more than one gene pair influence one characteristic or phenotypic trait.

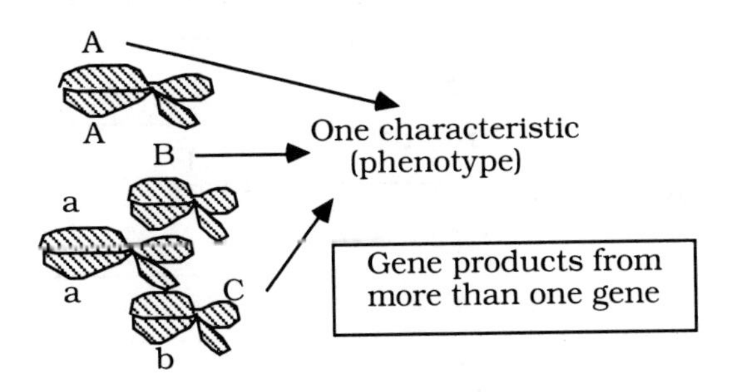

F4.2. Illustration of the symbolism associated with the wild type activity of a gene and several possible outcomes of the mutant state (*m/m*).

First, too much product (B) is made in the mutant state.

Second, too little product is made.

Third, no product is made.

The presence of the wild type allele (+) allows for the normal conversion of substrate A to product B.

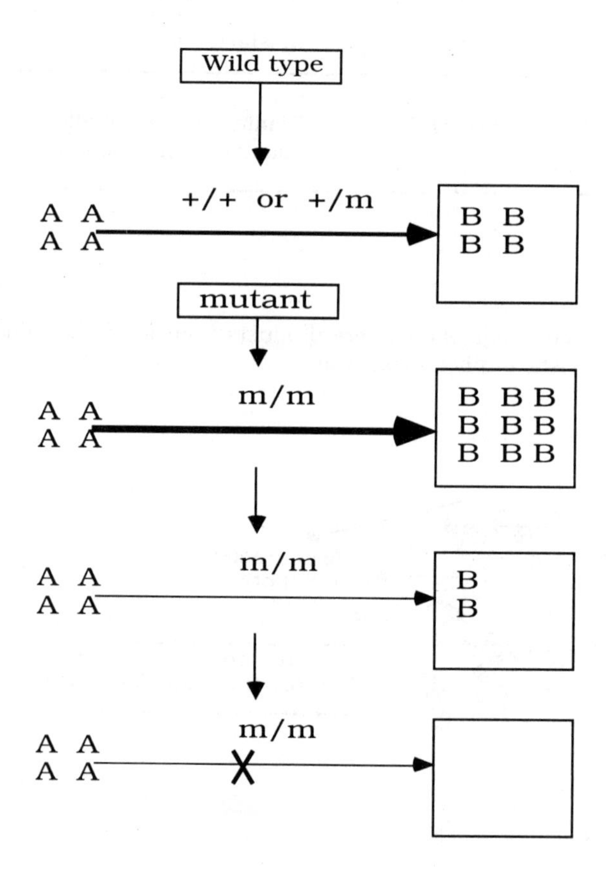

Solutions to Problems and Discussion Questions

1. In the first sentence of this problem, notice that there is one characteristic (coat color) and three phenotypes mentioned; red, white, or roan. The fact that roan is intermediate between red and white suggests that this may be a case of incomplete dominance, with roan being the intermediate. If that is the case then we should suspect a 1:2:1 phenotypic ratio in crosses of "roan to roan."

Looking at the data given, notice that a cross of the "extremes" (red X white) gives roan, suggesting its heterozygous nature and the homozygous nature of the parents. Seeing the 1:2:1 ratio in the offspring of

roan X roan

confirms the hypothesis of incomplete dominance as the mode of inheritance.

Symbolism:

AA = red

aa = white

Aa = roan

Crosses: It is important at this point that you not be fully dependent on writing out complete Punnett squares for each cross. Begin working these simple problems in your head.

AA	X	*AA*	—>	*AA*
aa	X	*aa*	—>	*aa*
AA	X	*aa*	—>	*Aa*
Aa	X	*Aa*	—>	1/4 *AA*; 2/4 *Aa*; 1/4 *aa*

2. *Incomplete dominance* can be viewed more as a quantitative phenomenon where the heterozygote is intermediate (approximately) between the limits set by the homozygotes. Pink is intermediate between red and white.

Codominance can be viewed in a more qualitative manner where both of the alleles in the heterozygote are expressed. For example in the AB blood group, both the I^A and I^B genes are expressed . There is no intermediate class which is part I^A and I^B.

3. In this problem remember that individuals with blood type B can have the genotype I^BI^B or I^BI^o and those with blood type A, genotypes I^AI^A or I^AI^o.

Male Parent: must be I^BI^o because the mother is I^oI^o and one inherits one homologue (therefore one allele) from each parent.

Female Parent: must be I^AI^o because the father is I^BI^o and one inherits one homologue (therefore one allele) from each parent. The father can not be I^BI^B and have a daughter of blood type A.

Offspring:

I^AI^o X I^BI^o

	I^B	I^o
I^A	$I^A I^B$(AB)	I^AI^o(A)
I^o	I^BI^o (B)	I^oI^o(O)

The ratio would be 1:1:1:1.

4. In *discontinuous variation* each phenotypic class is clearly discernible, while in *continuous variation* various genes interact to give a "blending" phenotype. *Epistasis* is a phenomenon involving gene interation where one gene or gene pair "masks" or inhibits expression of a non-allelic gene or gene pair(s) and as such is most likely grouped in the category of discontinuous inheritance.

5. Three independently assorting characteristics are being dealt with: flower color (incomplete dominance), flower shape (dominant/recessive), and plant height (dominant/recessive). Establish appropriate gene symbols:

Flower color:

RR = red ; *Rr* = pink; *rr* = white

Flower shape:

P = personate; *p* = peloric

Plant height:

D = tall; *d* = dwarf

(a)

RRPPDD X *rrppdd* —>

RrPpDd (pink, personate, tall)

(b) Use *components* of the forked line method as follows:

2/4 pink X 3/4 personate X 3/4 tall

= 18/64

6. There are two characteristics, flower color and flower shape. Because pink results from a cross of red and white, one would conclude that flower color is "monohybrid" with incomplete dominance.

In addition, because personate is seen in the F_1 when personate and peloric are crossed, personate must be dominant to peloric. Results from crosses (c) and (d) verify these conclusions. The appropriate symbols would be as follows:

Flower color:

RR = red; *Rr* = pink; *rr* = white

Flower shape:

P = personate; *p* = peloric

(a)

RRpp X *rrPP* ---> *RrPp*

(b)

RRPP X *rrpp* ---> *RrPp*

(c)

RrPp X *RRpp* ---> *RRPp*
RRpp
RrPp
Rrpp

(d)

RrPp X *rrpp* ---> *rrPp*
rrpp
RrPp
Rrpp

In the cross of the F_1 of (a) to the F_1 of (b), both of which are double heterozygotes, one would expect the following:

RrPp X *RrPp*

1/4 red
- 3/4 personate —> 3/16 red, personate
- 1/4 peloric —> 1/16 red, peloric

2/4 pink
- 3/4 personate —> 6/16 pink, personate
- 1/4 peloric —> 2/16 pink, peloric

1/4 white
- 3/4 personate —> 3/16 white, personate
- 1/4 peloric —> 1/16 white, peloric

7. Notice that the distribution of observed offspring fits a 9:3:4 ratio quite well. This suggests that two independently assorting gene pairs with epistasis are involved. Assign gene symbols in the usual manner:

A = pigment; *a* = pigmentless (colorless)

B = purple; *b* = red

AaBb X *AaBb*

produces

*A_B_* = purple
A_bb = red
*aaB_* = colorless
aabb = colorless

One may see this occurring in the following manner:

precursor —+-> cyanidin —+-> purple pigment
(colorless) *aa* (red) *bb*

8. This is a case of gene interaction (novel phenotypes) where the yellow and black types (double mutants) interact to give the cream phenotype and epistasis where the *cc* genotype produces albino.

(a)

AaBbCc —>

gray (*C* allows pigment)

(b)

A_B_Cc —>
gray (*C* allows pigment)

(c) Use the forked line method for this portion

```
                    1/2 Cc  —> 9/32 gray
         3/4 B_  <
                    1/2 cc  —> 9/32 albino
3/4 A_ <
                    1/2 Cc  —> 3/32 yellow
         1/4 bb  <
                    1/2 cc  —> 3/32 albino

                    1/2 Cc  —> 3/32 black
         3/4 B_  <
                    1/2 cc  —> 3/32 albino
1/4 aa <
                    1/2 Cc  —> 1/32 cream
         1/4 bb  <
                    1/2 cc  —> 1/32 albino
```

Combining the phenotypes gives (always count the proportions to see that they add up to 1.0):

16/32 albino

9/32 gray

3/32 yellow

3/32 black

1/32 cream

9. Treat each of the crosses as a series of monohybrid crosses, remembering that albino is epistatic to color and black and yellow interact to give cream.

(a) Since this is a 9:3:3:1 ratio with no albino phenotypes, the parents must each have been double heterozygotes and incapable of producing the *cc* genotype.

Genotypes:

AaBbCC X *AaBbCC*

or

AaBbCC X *AaBbCc*

Phenotypes: gray X gray

(b) Since there are no black offspring, there are no combinations in the parents which can produce *aa*. The 4/16 proportion indicates that the *C* locus is heterozygous in both parents. If the parents are as follows

AABbCc X *AaBbCc*

or

AABbCc X *AABbCc*

then the results would follow the pattern given.

Phenotypes: gray X gray

(c) Notice that 16/64 or 1/4 of the offspring are albino, therefore the parents are both heterozygous at the *C* locus. Second, notice that without considering the *C* locus, there is a 27:9:9:3 ratio which reduces to a 9:3:3:1 ratio. Given this information, the genotypes must be

AaBbCc X *AaBbCc.*

Phenotypes: gray X gray

10. In order to solve this problem one must first see the possible genotypes of the parents and the grandfathers. Since the gene is X-linked the cross will be symbolized with the X chromosomes.

RG = normal vision; rg = color-blind

Mother's father: X^{rg}/Y

Father's father: X^{rg}/Y

Mother: $X^{RG}X^{rg}$

Father: X^{RG}/Y

Notice that the mother must be heterozygous for the *rg* allele (being normal-visioned and having inherited an X^{rg} from her father) and the father, because he has normal vision, must be X^{RG}. The fact that the father's father is color-blind does not mean that the father will be color-blind. On the contrary, the father will inherit his X chromosome from his mother.

$X^{RG}X^{rg}$ X X^{RG}/Y —>

$X^{RG}X^{RG}$	= 1/4 daughter normal
$X^{RG}X^{rg}$	= 1/4 daughter normal
X^{RG}/Y	= 1/4 son normal
X^{rg}/Y	= 1/4 son color-blind

Looking at the distribution of offspring:

(a) 1/4

(b) 1/2

(c) 1/4

(d) zero

11. The mating is

$$X^{RG}X^{rg}; I^AI^O \quad X \quad X^{RG}Y; I^AI^O$$

Based on the son who is colorblind and blood type O, the mother must have been heterozygous for the *RG* locus and both parents must have had one copy of the I^O gene. The probability of having a female child is 1/2, that she has normal vision is 1.0 (because the father's X is normal) and 1/4 type O blood. The final product of the independent probabilities is

$$1/2 \text{ X } 1 \text{ X } 1/4 = 1/8$$

12. In seeing that the distribution of phenotypes in the F_1 is different when comparing males and females, it would be tempting to suggest that the gene is X-linked. However, given that the reciprocal cross gives identical results suggests that the gene is autosomal. Seeing the different distribution between males and females one might consider sex-influenced inheritance as a model and have males more likely to express mahogany and females more likely to express red. This situation is similar to pattern baldness in humans. Consider two alleles which are autosomal and let

RR = red, *Rr* = red in females,

Rr= mahogany in males,

rr = mahogany.

P_1:

female: *RR* (red) X male: *rr* (mahogany)

F_1:

Rr = females red; males mahogany

1/2 females (red)

1/2 males (mahogany)

F_2:

1/4 *RR*; 2/4 *Rr*; 1/4 *rr*

Because half of the offspring are males and half are females, one could, for clarity, rewrite the F_2 as:

	1/2 females	*1/2 males*
1/4 *RR*	1/8 red	1/8 red
2/4 *Rr*	2/8 red	2/8 mahogany
1/4 *rr*	1/8 mahogany	1/8 mahogany

13. This problem incorporates X-linked inheritance with mosaicism caused by X chromosome inactivation. The following symbolism is appropriate:

B = black coat color; *b* = yellow coat color

Female tortoise-shell = X^BX^b (mosaic)
Male black = X^BY

X^BX^b X X^BY —>

X^BX^B = female black

X^BX^b = female tortoise-shell

X^BY = male black

X^bY = male yellow

Normally there is no way for a tortoise-shell male to be produced, however with nondisjunction of the X chromosome in the female parent, thus producing a gamete containing the two X chromosomes, the coat color genes could be in the heterozygous state and mosaicism may result. The nondisjunctional event must have occurred in meiosis I. If such a female gamete is fertilized by a Y-bearing sperm, then an XXY tortoise-shell male could result.

14.

Symbolism:

Normal wing margins = sd^+;

scalloped = *sd*

(a)

P1:

$X^{sd}X^{sd}$ X X^+/Y —>

F_1:

1/2 X^+X^{sd} (female, normal)

1/2 X^{sd}/Y (male, scalloped)

F_2:

1/4 X^+X^{sd} (female, normal)

1/4 $X^{sd}X^{sd}$ (female, scalloped)

1/4 X^+/Y (male, normal)

1/4 X^{sd}/Y (male, scalloped)

(b)

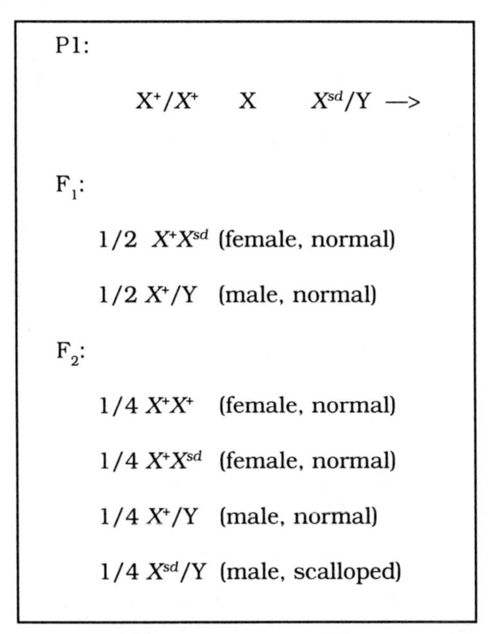
P1:

X^+/X^+ X X^{sd}/Y —>

F_1:

1/2 X^+X^{sd} (female, normal)

1/2 X^+/Y (male, normal)

F_2:

1/4 X^+X^+ (female, normal)

1/4 X^+X^{sd} (female, normal)

1/4 X^+/Y (male, normal)

1/4 X^{sd}/Y (male, scalloped)

If the *scalloped* gene were not X-linked, then all of the F_1 offspring would be wild (phenotypically) and a 3:1 ratio of normal to scalloped would occur in the F_2.

15. Assuming that the parents are homozygous, the crosses would be as follows. Notice that the X symbol may remain to remind us that the *sd* gene is on the X chromosome. It is extremely important that one account for both the mutant genes and each of their wild type alleles.

P_1:

$X^{sd}X^{sd}$; e^+/e^+ X X^+/Y; e/e —>

F_1:

1/2 X^+X^{sd}; e^+/e (female, normal)

1/2 X^{sd}/Y; e^+/e (male, scalloped)

F_2:

	X^+e^+	X^+e	$X^{sd}e^+$	$X^{sd}e$
$X^{sd}e^+$				
$X^{sd}e$				
Ye^+				
Ye				

Phenotypes:

3/16 normal females

3/16 normal males

1/16 ebony females

1/16 ebony males

3/16 scalloped females

3/16 scalloped males

1/16 scalloped, ebony females

1/16 scalloped, ebony males

Forked-line method:

P_1:

$X^{sd}X^{sd}$; e^+/e^+ X X^+/Y; e/e —>

F_1:

1/2 X^+X^{sd}; e^+/e (female, normal)

1/2 X^{sd}/Y; e^+/e (male, scalloped)

F_2:

	Wings	Color	
1/4	females, normal	3/4 normal	3/16
		1/4 ebony	1/16
1/4	females, scalloped	3/4 normal	3/16
		1/4 ebony	1/16
1/4	males, normal	3/4 normal	3/16
		1/4 ebony	1/16
1/4	males, scalloped	3/4 normal	3/16
		1/4 ebony	1/16

16. It is extremely important that one account for both the mutant genes and each of their wild type alleles.

(a)

P_1: X^vX^v; +/+ X X^+/Y; b^r/b^r —>

F_1:

1/2 X^+X^v; $+/b^r$ (female, normal)

1/2 X^v/Y; $+/b^r$ (male, vermilion)

F_2:

Eye color(X)	Eye color(autosomal)	
1/4 females, normal	3/4 normal	3/16
	1/4 brown	1/16
1/4 females, vermilion	3/4 normal	3/16
	1/4 brown	1/16
1/4 males, normal	3/4 normal	3/16
	1/4 brown	1/16
1/4 males, vermilion	3/4 normal	3/16
	1/4 brown	1/16

3/16 = females, normal

1/16 = females, brown eyes

3/16 = females, vermilion eyes

1/16 = females, white eyes

3/16 = males, normal

1/16 = males, brown eyes

3/16 = males, vermilion eyes

1/16 = males, white eyes

(b)

P_1: X^+X^+; b^r/b^r X X^v/Y; +/+ —>

F_1:

1/2 X^+X^v; $+/b^r$ (female, normal)
1/2 X^+/Y; $+/b^r$ (male, normal)

F_2:	Eye color(X)	Eye color(autosomal)
2/4	females, normal	3/4 normal
		1/4 brown
1/4	males, normal	3/4 normal
		1/4 brown
1/4	males, vermilion	3/4 normal
		1/4 brown

6/16 = females, normal

2/16 = females, brown eyes

3/16 = males, normal

1/16 = males, brown eyes

3/16 = males, vermilion eyes

1/16 = males, white eyes

(c)

P$_1$:

$X^vX^v;\ b^r/b^r$ X $X^+/Y;\ +/+$ —>

F$_1$:

1/2 $X^+X^v;\ +/b^r$ (female, normal)
1/2 $X^v/Y;\ +/b^r$ (male, vermilion)

F$_2$:	Eye color(X)	Eye color(autosomal)
1/4	females, normal	3/4 normal 1/4 brown
1/4	females, vermilion	3/4 normal 1/4 brown
1/4	males, normal	3/4 normal 1/4 brown
1/4	males, vermilion	3/4 normal 1/4 brown

3/16 = females, normal

1/16 = females, brown eyes

3/16 = females, vermilion eyes

1/16 = females, white eyes

3/16 = males, normal

1/16 = males, brown eyes

3/16 = males, vermilion eyes

1/16 = males, white eyes

17. Consider the solution in the following general manner. From cross number one, one should think about there being two gene loci which interact to give the phenotypes noted. Notice that two true-breeding sandy lines give red. This would happen if the following situation occurred:

> $s^1/s^1:S^2/S^2$ or $S^1/S^1:s^2/s^2$ homozygous strains each the sandy phenotype; also, as long as there is one upper case *S*, the sandy phenotype will prevail.

When crossed with each other they give the double heterozygotes:

> $S^1/s^1:\ s^2/S^2$ which gives the red phenotype because both S^1 and S^2 are present.

The white phenoytpe occurs when neither S^1 nor S^2 are present: $s^1/s^1:\ s^2/s^2$

Because crosses #1 and #4 are among double heterozygotes, independent assortment is occurring, and there are two ways to get the sandy phenotype, both F$_2$ results would be the following:

9 (red): 6(sandy): 1(white)

18. (a) Because the denominator in the ratios is 64 one would begin to consider that there are three independently assorting gene pairs. Because there are only two characteristics (eye color and croaking) however, one might consider two gene pairs interacting for one trait.

(b) Notice that there is a 48:16 (or 3:1) ratio of rib-it to knee-deep and a 36:16:12 (9:3:4) ratio of blue to green to purple eye color. Because of these relationships one would conclude that croaking is due to one gene pair while eye color is due to two gene pairs.

(c) Croaking: *R_* = rib-it; *rr* = knee-deep

*A_B_*	= blue eyed
A_bb	= purple
*aaB_* and *aabb*	= green

19. Parents: *AABBrr* X *AAbbRR*

F_1: *AABbRr*

F2: *AAB_R_* = blue-eyed, rib-it

AAB_rr = blue-eyed, knee-deep

*AAbbR_* = purple-eyed, rib-it

AAbbrr = purple-eyed, knee-deep

20. In doing these type of problems, take each characteristic individually, then build the complete genotypes. *AabbRr*

21. It is important to see that this problem involves multiple alleles, meaning that monohybrid type ratios are expected, and that there is an order of dominance which will allow certain alleles to be "hidden" in various heterozygotes. As with most genetics problems, one must look at the phenotypes of the offspring to assess the genotypes of the parents.

(a)

Parents: sepia X cream

Because both guinea pigs had albino parents, both are heterozygous for the c^a allele.

Cross:

c^kc^a X c^dc^a —>

2/4 sepia; 1/4 cream; 1/4 albino

(b)

Parents: sepia X cream

Because the sepia parent had an albino parent it must be c^kc^a. Because the cream guinea pig had two sepia parents

(c^kc^d X c^kc^d or c^kc^d X c^kc^a),

the cream parent could be c^dc^d or c^dc^a.

Crosses:

c^kc^a X c^dc^d —>

1/2 sepia; 1/2 cream*

*(if parents are assumed to be homozygous)

or

c^kc^a X c^dc^a —>

1/2 sepia; 1/4 cream; 1/4 albino

(c)

Parents: sepia X cream

Because the sepia guinea pig had two full color parents which could be

Cc^k, Cc^d, or Cc^a

(not *CC* because sepia could not be produced),

its genotype could be

c^kc^k, c^kc^d, or c^kc^a.

Because the cream guinea pig had two sepia parents

(c^kc^d X c^kc^d or c^kc^d X c^kc^a),

the cream parent could be c^dc^d or c^dc^a.

Crosses:

c^kc^k X c^dc^d —> all sepia

c^kc^k X c^dc^a—> all sepia

c^kc^d X c^dc^d —> 1/2 sepia; 1/2 cream

c^kc^d X c^dc^a —> 1/2 sepia; 1/2 cream

c^kc^a X c^dc^d —> 1/2 sepia; 1/2 cream

c^kc^a X c^dc^a —>

1/2 sepia; 1/4 cream; 1/4 albino

(d)

Parents: sepia X cream

Because the sepia parent had a full color parent and an albino parent

(Cc^k X c^ac^a),

it must be c^kc^a. The cream parent had two full color parents which could be Cc^d or Cc^a; therefore it could be c^dc^d or c^dc^a.

Crosses:

c^kc^a X c^dc^d —> 1/2 sepia; 1/2 cream

c^kc^a X c^dc^a —>

1/2 sepia; 1/4 cream; 1/4 albino

22. (a)

(i) *BbYy* = green parents

(ii, iii)

*B_Y_* = green progeny (9/16)

B_yy = blue progeny (3/16)

*bbY_* = yellow progeny (3/16)

bbyy = albino progeny (1/16)

(b) *BByy* X *bbYY* or *BBYY* X *bbyy*

23. (a,b) In looking at the pedigrees, one can see that the condition cannot be dominant because it appears in the offspring (II-3 and II-4) and not the parents in the first two cases. The condition is therefore *recessive*. In the second cross, note that the father is not shaded yet the daughter (II-4) is. If the condition is recessive, then it must also be *autosomal*.

(c) II-1= *AA* or *Aa*

II-6= *AA* or *Aa*

II-9= *Aa*

Quantitative Genetics

Vocabulary: Organization and Listing of Terms

Structures and Substances

- DDT

Processes/Methods

- Discontinuous traits
- Continuous traits
  - quantitative inheritance
  - biometry
  - multiple-factor (gene) hypothesis
  - additive alleles
    - 1:4:6:4:1
    - $1/4^n$ (n = number of gene pairs)
    - 2n + 1 rule
  - polygenic
    - phenotypic flexibility
    - interaction with the environment
      - resistance to DDT (*Drosophila*)
    - RFLP analysis
- Statistical analysis
  - descriptive summary
  - statistical inference
  - statistics
  - parameters
    - mean
      - central tendency
    - variance
    - standard deviation
    - standard error of the mean
- Heredity vs. environment
  - broad heritability
  - narrow heritability
  - twin studies
    - monozygotic (identical) twins
    - dizygotic (fraternal) twins
      - concordant vs. discordant

Mapping

- quantitative trait loci (QTL)
- non-random segregation with RFLP

Concepts

Quantitative inheritance

Heritability

Phenotypic variation

Heredity vs. environment

- inbred strains
- isogenic
- heritability index or ratio (H^2)
    - broad heritability
        - environmental variance
        - genetic variance
        - interaction
    - narrow heritability
        - dominance variance
        - additive variance
        - interactive variance

Concordance

Discordance

Artificial selection

Twin studies

Mapping

Solutions to Problems and Discussion Questions

1. In *discontinuous* variation the influences of each gene pair are not additive and more typical Mendelian ratios such as 9:3:3:1 and 3:1 result. In *continuous* variation, different gene pairs interact (usually additively) to produce a phenotype which is less "stepwise" in distribution. Inheritance of a quantitative nature follows a more continuous form.

2. (a) *Polygenes* are those genes which are involved in determining continuously varying or multiple factor traits.

(b) *Additive alleles* are those alleles which account for the hereditary influence on the phenotype in an additive way.

(c) The *multiple factor hypothesis* suggested that many factors or genes contribute to the phenotype in a cumulative or quantitative way.

(d) *Monozygotic twins* are derived from a single fertilized egg and are thus genetically identical to each other. They provide a method for determining the influence of genetics and environment on certain traits. *Dizygotic twins* arise from two eggs fertilized by two sperm cells. They have the same genetic relationship as siblings.

The role of genetics and the role of the environment can be studied by comparing the expression of traits in monozygotic and dizygotic twins. The higher concordance value for monozygotic twins as compared to the value for dizygotic twins indicates a significant genetic component for a given trait.

(e) *Concordance* refers to the frequency with which both members of a twin pair express a given trait. *Discordance* refers to the frequency at which one twin expresses a trait while the other does not. A comparison of concordance (and discordance) frequencies can provide information on the genetic and/or environmental influence on a given trait.

(f) *Heritability* is a measure of the degree to which the phenotypic variation of a given trait is due to genetic factors. A high heritability indicates that genetic factors are major contributers to phenotypic variation while environmental factors have little impact.

3. If you add the numbers given for the ratio, you obtain the value of 16, which is indicative of a dihybrid cross. The distribution is that of a dihybrid cross with additive effects.

(a) Because a dihybrid result has been identified, there are two loci involved in the production of color. There are two alleles at each locus for a total of four alleles.

(b) Because the description of red, medium-red, etc., gives us no indication of a *quantity* of color in any form of units, we would not be able to actually quantify a unit amount for each change in color. We can say that each gene (additive allele) provides an equal *unit* amount to the phenotype and the colors differ from each other in multiples of that unit amount.

(c) The genotypes are as follows:

1/16	= dark red	=	*AABB*
4/16	= medium-dark red	=	2*AABb* 2*AaBB*
6/16	= medium red	=	*AAbb* 4*AaBb* *aaBB*
4/16	= light red	=	2*aaBb* 2*Aabb*
1/16	= white	=	*aabb*

(d) F_1 = all light red

F_2 = 1/4 medium red
2/4 light red
1/4 white

4. (a) It *is possible* that two parents of moderate height can produce offspring that are much taller or shorter than either parent because segregation can produce a variety of gametes, therefore offspring as illustrated below:

rrSsTtuu X *RrSsTtUu*
(moderate) (moderate)

Offspring from this cross can range from very tall *RrSSTTUu* (12 "tall" units) to very short *rrssttuu* (8 "small" units).

(b) If the individual with a minimum height, *rrssttuu*, is married to an individual of intermediate height *RrSsTtUu*, the offspring can be no taller than the height of the tallest parent. Notice that there is no way of having more than four dominant alleles in the offspring.

5. As you read this question, notice that the strains are inbred, therefore homozygous, and that approximately 1/250 represent the shortest and tallest groups in the F_2 generation.

(a, b) Referring to the text, see that where four gene pairs act additively, the proportion of one of the extreme phenotypes to the total number of offspring is 1/256 (add the numbers in each phenotypic class). The same may be said for the other extreme type. The extreme types in this problem are the 12cm and 36cm plants. From this observation one would suggest that there are four gene pairs involved.

(c) If there are four gene pairs, there are nine (2n+1) phenotypic categories and eight increments between these categories. Since there is a difference of 24cm between the extremes, 24cm/8 = 3cm for each increment (each of the additive alleles).

(d) A typical F_1 cross which produces a "typical" F_2 distribution would be where all gene pairs are heterozygous (*AaBbCcDd*), independently assorting, and additive. There are many possible sets of parents which would give an F_1 of this type. The limitation is that each parent has genotypes which give a height of 24cm as stated in the problem. Because the parents are inbred, it is expected that they are fully homozygous.

An example:

AABBccdd X *aabbCCDD*

(e) Since the *aabbccdd* genotype gives a height of 12cm and each upper-case allele adds 3cm to the height, there are many possibilities for an 18 cm plant:

AAbbccdd,

AaBbccdd,

aaBbCcdd, etc.

Any plant with seven upper-case letters will be 33cm tall:

AABBCCDd,

AABBCcDD,

AABbCCDD, for examples.

6. (a) There is a fairly continuous range of "quantitative" phenotypes in the F_2 and an F_1 which is between the phenotypes of the two parents; therefore, one can conclude that some phenotypic blending is occurring which is probably the result of several gene pairs acting in an additive fashion. Because the extreme phenotypes (6cm and 30cm) each represent 1/64 of the total, it is likely that there are three gene pairs in this cross.

Remember, trihybrid crosses which show independent assortment of genes have a denominator (4^3) of 64 in ratios. Also, the fact that there are seven categories of phenotypes, which, because of the relationship 2n+1 = 7, would give the number of gene pairs (n) of 3. The genotypes of the parents would be combinations of alleles which would produce a 6cm (*aabbcc*) tail and a 30cm (*AABBCC*) tail while the 18cm offspring would have a genotype of *AaBbCc*.

(b) A mating of an *AaBbCc* (for example) pig with the 6cm *aabbcc* pig would result in the following offspring:

Gametes (18cm tail)	*Gamete (6cm tail)*	*Offspring*	
ABC		*AaBbCc*	(18cm)
ABc		*AaBbcc*	(14cm)
AbC		*AabbCc*	(14cm)
Abc	*abc*	*Aabbcc*	(10cm)
aBC		*aaBbCc*	(14cm)
aBc		*aaBbcc*	(10cm)
abC		*aabbCc*	(10cm)
abc		*aabbcc*	(6cm)

In this example, a 1:3:3:1 ratio is the result. However, had a different 18cm tailed-pig been selected, a different ratio would occur:

AABbcc X aabbcc

Gametes (18cm tail)	*Gamete (6cm tail)*	*Offspring*
ABc	*abc*	*AaBbcc* (14cm)
Abc		*Aabbcc* (10cm)

7. For height, notice that average differences between MZ twins reared together (1.7 cm) and those MZ twins reared apart (1.8 cm) are similar (meaning little environmental influence) and considerably less than differences of DZ twins (4.4 cm) or sibs (4.5) reared together. These data indicate that genetics plays a major role in determining height.

However, for weight, notice that MZ twins reared together have a much smaller (1.9 kg) difference than MZ twins reared apart, indicating that the environment has a considerable impact on weight. By comparing the weight differences of MZ twins reared apart with DZ twins and sibs reared together one can conclude that the environment has almost as much an influence on weight as genetics.

8. *Monozygotic twins* are derived from a single fertilized egg and are thus genetically identical to each other. They provide a method for determining the influence of genetics and environment on certain traits. *Dizygotic twins* arise from two eggs fertilized by two sperm cells. They have the same genetic relationship as siblings.

The role of genetics and the role of the environment can be studied by comparing the expression of traits in monozygotic and dizygotic twins. The higher concordance value for monozygotic twins as compared to the value for dizygotic twins indicates a significant genetic component for a given trait.

9. Many traits, especially those which we view as quantitative are likely to be determined by a polygenic mode. The following are some common examples: height, general body structure, skin color, and perhaps most common behavioral traits including intelligence.

10. At first glance, this problem looks as if it will be an arithmetic headache, however, the problem can be simplified.

(a) The mean is computed by adding the measurements of all of the individuals, then dividing by the number of individuals. In this case there are 760 corn plants. To keep from having to add 760 numbers, merely multiply each height group by the number of individuals in each group. Add all the products then divide by n (760). This gives a value for the mean of 140 cm.

(b) For the variance, use the formula given below (as in the text):

$$s^2 = V = n\Sigma f(x^2) - (\Sigma fx)^2 / n(n - 1)$$

To simplify the calculations, determine the square of each height group (100 cm for example) then multiply the value by the number in each group.

For the first group (100 cm) we would have:

100 X 100 X 20= 200000

The rest of the groups are as follows:

110 X 110 X 60 = 726000
120 X 120 X 90 = 1296000
130 X 130 X 130 = 2197000
140 X 140 X 180 = 3528000
150 X 150 X 120 = 2700000
160 X 160 X 70 = 1792000
170 X 170 X 50 = 1445000
180 X 180 X 40 = 1296000

= 15180000

Now, the mean squared, multiplied by n is as follows:

140 X 140 X 760 = 14896000

Completing the calculations gives the following:

(15180000 - 14896000)/759

= 284000/759

= 374.18

(c) The *standard deviation* is the square root of the variance or 19.34.

(d) The *standard error of the mean* is the standard deviation divided by the square root of n, or about 0.70.

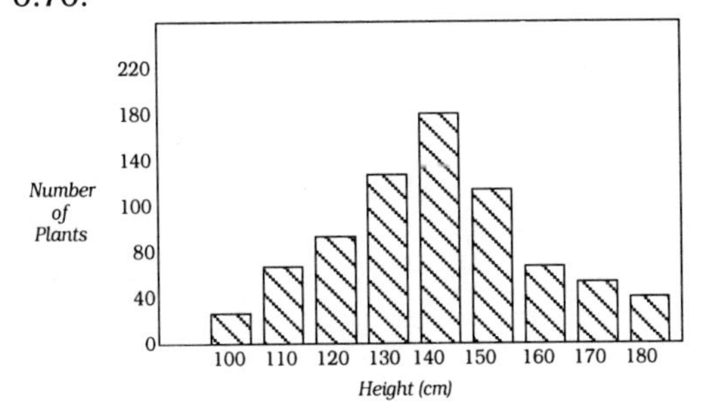

The plot approximates a normal distribution. Variation is continuous.

11. For a trait that is quantitatively measured, the relative importance of genetic *versus* environmental factors may be formally assessed by examining the heritability index (H^2 or broad heritability). In animal and plant breeding, a measure of potential response to selection based on additive variance and dominance variance is termed narrow heritability (h^2). A relatively high narrow heritability is a prediction of the impact selection may have in altering an initial randomly breeding population.

12. The formula for estimating heritability is

$$H^2 = V_G/V_P \text{ where } V_G \text{ and } V_P$$

are the genetic and phenotypic components of variation, respectively. The main issue in this question is obtaining some estimate of two components of phenotypic variation: genetic and environmental.

V_P is the combination of genetic and environmental variance. Because the two parental strains are inbred, they are assumed to be homozygous and the variance of 4.2 and 3.8 considered to be the result of environmental influences. The average of these two values is 4.0. The F_1 is also genetically homogeneous and gives us an additional estimation of the environmental factors.

By averaging with the parents

[(4.0 + 5.6)/2 = 4.8]

we obtain a relatively good idea of environmental impact on the phenotype. The phenotypic variance in the F_2 is the sum of the genetic (V_G) and environmental (V_E) components. We have estimated the environmental input as 4.8, so 10.3 minus 4.8, gives us an estimate of (V_G) which is 5.5. Heritability then becomes 5.5/10.3 or 0.53. This value, when viewed in percentage form indicates that about 53% of the variation in plant height is due to genetic influences.

13. **(a)** For Vitamin A

$h_A^2 = V_A/V_P = V_A/(V_E + V_A + V_D) = 0.097$

For Cholesterol

$h_A^2 = 0.223$

(b) Cholesterol content should be influenced to a greater extent by selection.

14. $h^2 = (7.5 - 8.5/6.0 - 8.5) = 0.4$

Selection will have little relative influence on olfactory learning in *Drosophila*.

15. $h^2 = 0.3 = (M_2 - 60/80 - 60)$

$M_2 = 66$ grams

Linkage and Chromosome Mapping

Vocabulary: Organization and Listing of Terms

Structures and Substances

Drosophila

Neurospora

Chlamydomonas

- tetrad

Zea mays

Chiasma

Heterokaryon

Synkaryon

Ascospores

Ascus (pl. asci)

Bromodeoxyuridine (BUdR)

Processes/Methods

Linkage

- crossing over
  - crossover gametes (recombinant)
  - parental gametes

recombination

- reciprocal classes

complete

- parental (noncrossover gametes)

incomplete

chiasmata (chiasma)

three-point mapping

- product rule (multiple crossovers)
  - non-crossovers (NCO)
  - single crossovers (SCO)
  - double crossovers (DCO)

Determining gene sequence

Gene to centromere mapping

- first division segregation
- second division segregation

$$\frac{1/2(\text{second division segregation})}{\text{total asci scored}}$$

Cytological evidence (crossing over)

- cytological markers

Somatic cell hybridization

- random loss of human chromosomes
- synteny testing

Haploid organisms

- tetrad analysis

Sister chromatid exchange

- bromodeoxyuridine (BUdR)
- Bloom syndrome

Mendel and linkage

- independent assortment

Concepts

Linked genes (linkage groups)

- arrangement (F6.1)

Chromosome maps

- map unit (% recombination)
- centimorgan (cM)
- 50% maximum
- influence of sex of organism

Interference

- coefficient of coincidence
- expected frequency of DCO
- observed frequency of DCO
    - positive interference
    - negative interference

Generation of variation

F6.1. Illustration of two typical configurations of two heterozygous gene pairs. Understanding of such arrangements is key to doing linkage problems.

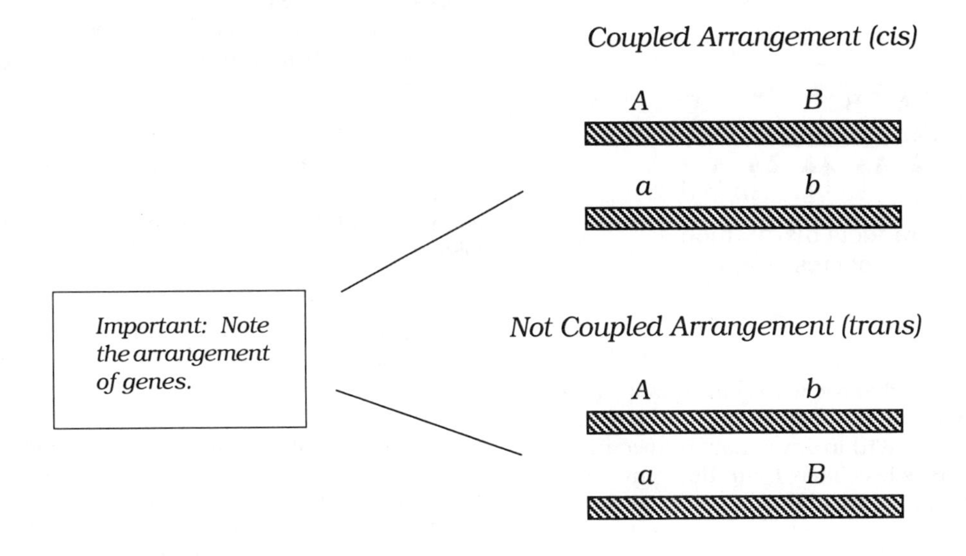

Solutions to Problems and Discussion Questions

1. With some qualification, one can say that crossing over is randomly distributed over the length of the chromosome. Two loci which are far apart are more likely to have a crossover between them than two loci that are close together.

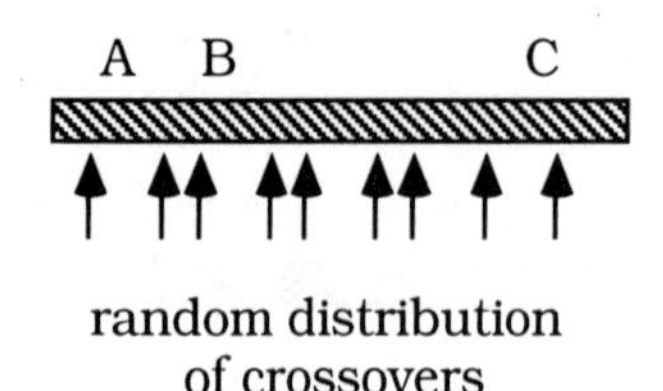

random distribution of crossovers

2. As mentioned above with some qualifications, crossovers occur randomly along the lengths of chromosomes. Within any region, the occurrence of two events is less likely than the occurrence of one event. If the probability of one event is

$$1/X,$$

the probability of two events occurring at the same time will be

$$1/X^2.$$

3. Each cross must be set-up in such a way as to reveal crossovers because it is on the basis of crossover frequency that genetic maps are developed. It is necessary that genetic heterogeneity exist so that different arrangements of genes, generated by crossing over, can be distinguished.

The organism which is heterozygous must be the sex in which crossing over occurs. In other words, it would be useless to map genes in *Drosophila* if the male parent is the heterozygote since crossing over is not typical in *Drosophila* males.

Lastly, the cross must be set-up so that the phenotypes of the offspring readily reveal their genotypes. The best arrangement is one where a fully heterozygous organism is crossed with an organism which is fully recessive for the genes being mapped.

4. Since the distance between *dp* and *ap* is greatest, they must be on the "outside" and *cl* must be in the middle. The genetic map would be as *follows:*

dp—*cl*————————*ap*
3 mu. 39 mu.

5. In looking at this problem one can immediately conclude that the two loci (kernel color and plant color) are linked because the test cross progeny occur in a ratio other than 1:1:1:1 (and epistasis does not appear because all phenotypes expected are present). The question is whether the arrangement in the parents is *coupled* (see F6.1)

RY/ry X *ry/ry*

or *not coupled* (see F6.1)

Ry/rY X *ry/ry*

Notice that the most frequent phenotypes in the offspring, the parentals, are colored, green (88) and colorless, yellow (92). This indicates that the heterozygous parent in the test cross is *coupled*

$$RY/ry \quad X \quad ry/ry$$

with the two dominant genes on one chromosome and the two recessives on the homologue (F6.1). Seeing that there are 20 crossover progeny among the 200, or 20/200, the map distance would be 10 map units (20/200 X 100 to convert to percentages) between the *R* and *Y* loci.

6. Since there is no indication as to the configuration of the *P* and *Z* genes (*coupled* or *not coupled*) in the parent, one must look at the percentages in the offspring. Notice that the most frequent classes are *PZ* and *pz*. These classes represent the parental (non-crossover) groups which indicates that the original parental arrangement in the test cross was

$$PZ/pz \quad X \quad pz/pz$$

Adding the crossover percentages together

(6.9 + 7.1) gives 14%

which would be the map distance between the two genes.

7.

	female A:	*female B:*	*Frequency:*
NCO	3, 4	7, 8	first
SCO	1, 2	3, 4	second
SCO	7, 8	5, 6	third
DCO	6, 5	1, 2	fourth

The single crossover classes which represent crossovers between the genes which are closer together (*d-b*) would occur less frequently than the classes of crossovers between more distant genes (*b-c*).

8. For two reasons, it is clear that the genes are in the *coupled* configuration (see F6.1) in the F_1 female. First, a completely homozygous female was mated to a wild type male and second, the phenotypes of the offspring indicate the following parental classes

sc s v and + + +

(a)

P_1: *sc s v /sc s v* X + + +/Y

F_1: + + +/*sc s v* X *sc s v*/Y

(b) For determining the sequence of genes, examine the parental classes and compare the arrangement with the double crossover (least frequent) classes. Notice that the *v* gene "switches places" between the two groups (parentals and double crossovers). The gene which switches places is in the middle.

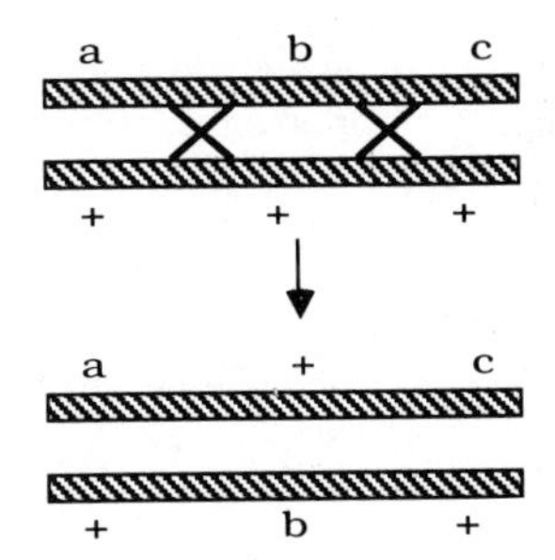

The map distances are determined by first writing the proper arrangement and sequence of genes, then computing the distances between each set of genes.

$$\frac{sc \quad v \quad s}{+ \quad + \quad +}$$

$$sc - v = \frac{150 + 156 + 10 + 14}{1000} \times 100 = 33\% \text{ (map units)}$$

$$v - s = \frac{46 + 30 + 10 + 14}{1000} \times 100 = 10\% \text{ (map units)}$$

Double crossovers are always added into each crossover group because they represent a crossover in each region.

sc——*v*——*s*
33 10

(c) The coefficient of coincidence=

$$\frac{\text{observed freq. double C/O}}{\text{expected freq. double C/O}}$$

$$= \frac{(14 + 10)/1000}{.33 \text{ X } .1}$$

$$= \frac{.024}{.033}$$

$$= .727$$

which indicates that there were fewer double crossovers than expected, therefore positive chromosomal interference is present.

9. (a) The cross will be as follows. Represent the *Dichete* gene as an upper-case letter because it is dominant.

P_1:

D + +/ + + + X *+ e p/+ e p*

F_1:

D + +/+ e p X *+ e p/+ e p*

F_2:

D + +/+ e p	Dichete
+ e p /+ e p	ebony, pink
D e +/+ e p	Dichete, ebony
+ + p/+ e p	pink
D + p/+ e p	Dichete, pink
+ e +/+ e p	ebony
D e p/+ e p	Dichete, ebony, pink
+ + +/+ e p	wild type

(b) Determine which gene is in the middle by comparing the parental classes with the double crossover classes. Notice that the *pink* gene "switches places" between the two groups (parentals and double crossovers). The gene which switches places is in the middle. So rewriting the sequence of genes with the correct arrangement gives the following:

F_1:

D + +/+ p e X *+ p e/+ p e*

Distances: remember to add in the double crossover classes

$$D - p = \frac{12 + 13 + 2 + 3}{1000} \text{ X } 100$$

$$= 3.0 \text{ map units}$$

$$p - e = \frac{84 + 96 + 2 + 3}{1000} \text{ X } 100$$

$$= 18.5 \text{ map units}$$

10. The map distance of a gene to the centromere in *Neurospora* is determined by dividing the percentage of second division asci (tetrads) by two. Patterns other than *BBbb* or *bbBB* are "second division" as discussed in the text and represent a crossover between the gene in question and the centromere. In the data given, the percentage of second division segregation is 20/100 or 20%. Dividing by 2 (because only two of the four chromatids are involved in any single crossover event) gives 10 map units.

11. First make a drawing with the genes placed on the homologous chromosomes as follows:

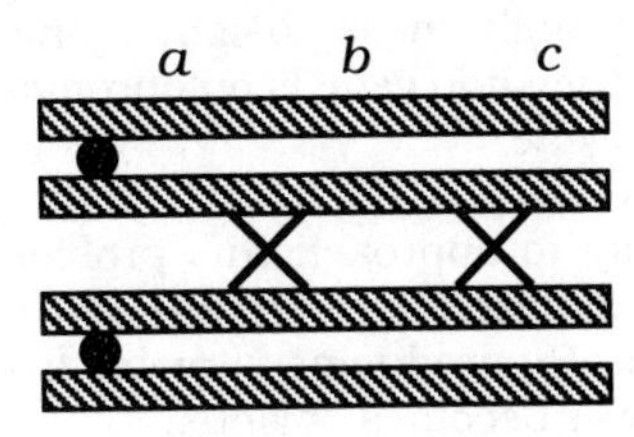

Realize that there are four chromatids in each tetrad and a single crossover involves only two of the four chromatids. The non-involved chromatids must be added to the non-crossover classes. Do all the crossover classes first, then add up the non-crossover chromatids.

For example, in the first crossover class (20 between *a* and *b*) notice that there will be 40 chromatids which were not involved in the crossover. These 40 must be added to the *abc* and +++ classes.

a b c	=	168
+ + +	=	168
a + +	=	20
+ *b c*	=	20
+ + *c*	=	10
a b +	=	10
+ *b* +	=	2
a + *c*	=	2

The map distances would be computed as follows:

$$a - b = \frac{20 + 20 + 2 + 2}{400} \times 100$$

$$= 11 \text{ map units}$$

$$b - c = \frac{10 + 10 + 2 + 2}{400} \times 100$$

$$= 6 \text{ map units}$$

12. Since *Stubble* is a dominant mutation (and homozygous lethal) one can determine whether it is heterozygous (*Sb*/+) or homozygous wild type (+/+). One would use the typical test cross arrangement with the *curled* gene so the arrangement would be

+ *cu*/ + *cu*

13. This set-up involves an F_1 in which the fully heterozygous female has the genes *y* and *w* in *coupled* and *ct* *not coupled.* The arrangement for the cross is therefore:

(a) *y w* +/+ + *ct* X *y w* +/Y

It is important at this point to determine the gene sequence. Examine the parental classes and compare the arrangement with the double crossover (least frequent) classes. Notice that the *w* gene "switches places" between the two groups (parentals and double crossovers). The gene which switches places is in the middle. Therefore the arrangement as written above is correct.

(b)

$$y - w = \frac{9 + 6 + 0 + 0}{1000} \times 100$$

$$= 1.5 \text{ map units}$$

$$w - ct = \frac{90 + 95 + 0 + 0}{1000} \times 100$$

$$= 18.5 \text{ map units}$$

y —— *w* —————— *ct*
0.0 1.5 20.0

(c) There were

.185 X .015 X 1000 = 2.775

double crossovers expected.

(d) Because the cross to the F_1 males included the normal (wild type) gene for *cut wings* it would not be possible to unequivocally determine the genotypes from the F_2 phenotypes for all classes.

14. This problem can be approached by looking for the most distant loci (*adp* and *b*) then filling in the intermediate loci. In this case the map for parts **(a)** and **(b)** is the following:

d	b	pr	vg	c	adp
31	48	54	67	75	83

Map Units

The expected map units between *d* and *c* would be 44, *d* and *vg* would be 36, and *d* and *adp* 52, however because there is a theoretical maximum of 50 map units possible between two loci in any one cross, that distance would be below the 52 determined by simple subtraction.

15. **(a)** $2^n = 8$

(b) 2 (no crossing over)

(c) *A* and *B* loci are tightly linked and there are 10 map units between either *A* or *B* and *C*.

16. **(a)** There are several ways to think through this problem. Remember that there is no crossing over in *Drosophila* males, therefore any gene on the same chromosome will be completely linked to any other gene on the same chromosome. Since you can get *pink* by itself, *short* can not be completely linked to it. This leaves linkage to *black* on the second chromosome, the 4th chromosome, or the X chromosome. Since the distribution of phenotypes in males and females is essentially the same, the gene can not be X-linked. In addition, the F_1 males were wild and if the *short* gene is on the X, the F_1 males would be short.

It is also reasonable to state that the gene can not be on the 4th chromosome because there would be eight phenotypic classes (independent assortment of three genes) instead of the four observed. Through these insights, one could conclude that the *short* gene is on chromosome 2 with the *black* gene.

Another way to approach this problem is to make three chromosomal configurations possible in the F_1 male. By producing gametes from this male, the answer becomes obvious.

Case A	Case B	Case C
$\frac{p\ b \quad sh}{+\ + \quad +}$	$\frac{p\ sh \quad b}{+\ + \quad +}$	$\frac{b\ sh \quad p}{+\ + \quad +}$

Develop the gametes from Case C and cross them out to the completely recessive triple mutant. You will get the results in the table.

(b) The parental cross is now the following:

Females: $\frac{b\ sh \quad p}{+\ + \quad +}$ X Males: $\frac{b\ sh \quad p}{b\ sh \quad p}$

The new gametes resulting from crossing over in the female would be *b* + and + *sh*. Since the gene *p* is assorting independently, it is not important in this discussion. Because 15% of the offspring now contain these recombinant chromatids, the map distance between the two genes must be 15.

Non-Mendelian Inheritance

Vocabulary: Organization and Listing of Terms

Structures and Substances

Kynurenine

Tryptophan

Heterokaryon

Hyphae

Mycelia

Paramecin

Kappa

Plasmids, episomes

F factors

- Hfr

R factors (RTF)

- *Shigella*

Col factor

- colicins

Processes/Methods

Extrachromosomal (Extranuclear) influence

- maternal influence (effect)
  - transitory
  - *Ephestia kuehniella* (*A*, *a*)
  - *Limnaea peregra* (*D*, *d*)
    - dextral, sinistral
      - injection experiments
- organelle heredity
  - chloroplast DNA
    - *Mirabilis jalapa*
    - Maize (*iojap*)
    - *Chlamydomonas reinhardi* (s^r)
  - mitochondrial DNA
    - *Neurospora crassa* (*poky*)
      - suppressive

Saccharomyces cerevisiae (*petite*)

segregational

neutral

suppressive

Humans

heteroplasmy

myoclonic epilepsy

(MERRF)

Leber's hereditary optic

Kearns-Sayre syndrome

infectious heredity

Paramecium aurelia

kappa

conjugation

autogamy

Drosophila

CO_2 sensitivity

sigma

D. bifasciata, D. willistoni

sex ratio

protozoan

Plasmids

Conjugations

Resistance transfer

Concepts

Extrachromosomal influence (F7.1)

Non-Mendelian patterns of inheritance (F7.1)

Plasmid biology

F7.1. Illustration of the common pattern seen in many cases of extrachromosomal inheritance. The condition of the egg parent has a stronger influence on the phenotype of the offspring than the sperm/pollen parent. Therefore reciprocal crosses give different results.

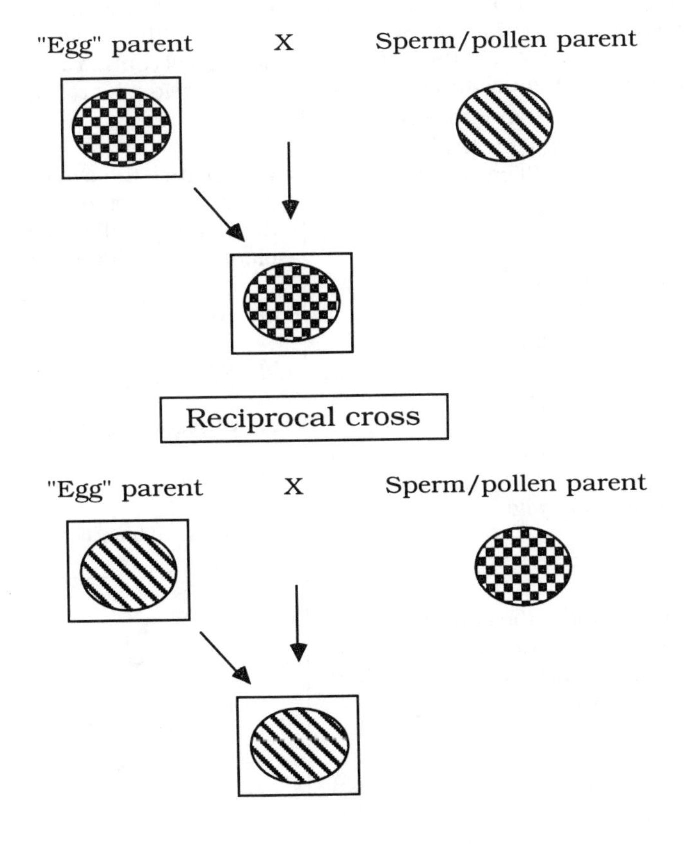

Solutions to Problems and Discussion Questions

1. In cases of extrachromosomal inheritance, the phenotype is determined by the genetic (maternal effect) or cytoplasmic (organelle or infectious) condition of the parent which contributes the bulk of the cytoplasm to the offspring. In most cases, the maternal parent provides the basis for the cytoplasmic inheritance.

The pattern of inheritance is more often from one parent to the offspring. One does not see both parents contributing to the characteristics of the offspring as is the case with Mendelian (chromosomal) forms of inheritance. Standard Mendelian ratios (3:1) are usually not present. In general, the results of reciprocal crosses differ. See F7.1.

Female mutant X male wild

all offspring mutant

Female wild X male mutant

all offspring wild

In sex-linked inheritance, the pattern is often from grandfather through carrier mother to son. Patterns of extrachromosomal inheritance are often not influenced by the sex of the individual.

2. The case with *Limnaea* involves a maternal effect in which the *genotype* of the mother influences the *phenotype* of the *immediate* offspring in a non-Mendelian manner. Notice that in the above statement, it is the maternal genotype which determines the phenoytpe of the offspring, regardless of its own genotype.

Since both of the parents are *Dd*, the parent contributing the eggs must be *Dd*. Therefore, all of the offspring must have the phenotype of the mother's genotype, which is dextral.

3. The mt^+ strain (resistant for the nuclear and chloroplast genes) contributes the "cytoplasmic" component of streptomycin resistance which would negate any contribution from the mt^- strain. Therefore, all the offspring will have the streptomycin resistance phenotype. In the reciprocal cross, with the mt^+ strain being streptomycin sensitive, the nuclear genome of half of the offspring will contain a streptomycin resistance gene and therefore be resistant.

4. Because the ovule source furnishes the cytoplasm to the embryo and thus the chloroplasts, the offspring will have the same phenotype as the plant providing the ovule.

a) green

b) white

c) variegated (patches of white and green)

d) green

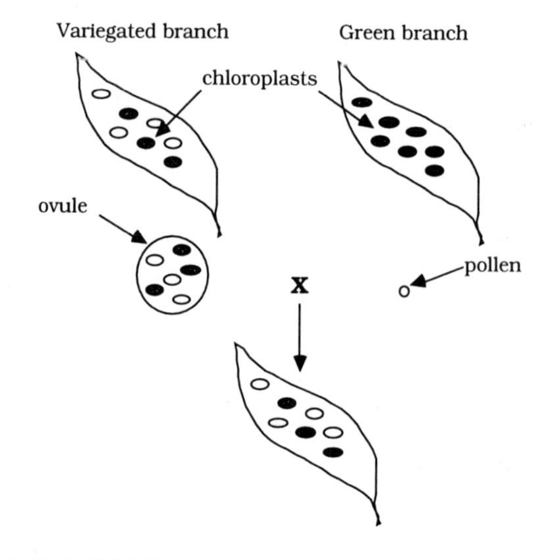

5. As with any description of dominance, one looks to the phenotype of the diploid heterozygote. In this problem, the heterozygote is of normal phenotype, therefore the *petite* gene is recessive.

6. Examine the *Essentials* text and notice that the inheritance patterns for the two, *segregational* and *neutral*, are quite different. The segregational mode is dependent on nuclear genes while that of the neutral type is dependent on cytoplasmic influences, namely mitochondria. If the two are crossed as stated in the problem, then one would expect, in the diploid zygote, the *segregational* allele to be "covered" by normal alleles from the neutral strain.

On the other hand, as the nuclear genes are again "exposed" in the haploid state of the ascospores, one would expect a 1:1 ratio of normals to petites. The petite phenoytpe is caused by the nuclear, *segregational* gene.

7. In providing the answers to this question, remember that a *Paramecium* may be sensitive and carry the *K* gene. These are organisms which did not obtain kappa particles. There is a question as to whether cytoplasmic exchange has occurred, however, seeing the results in cross (b) one can assume that cytoplasmic exchange has occurred. In addition, one can assume that only the exconjugants are being described in the offspring. Under those conditions, the parental genotypes could be the following:

(a) *Kk* X *kk*

(b) any case where there is no *kk* such as:

KK X *KK*

or

KK X *kk*

(c) *Kk* X *Kk*

8. (a) There are many similarities among mitochondrial, chloroplast, and prokaryotic molecular systems. It is likely that mitochondria and chloroplasts evolved from bacteria in a symbiotic relationship, therefore it is not surprising that certain antibiotics which influence bacteria will also influence all mitochondria and chloroplasts.

(b) Clearly, the mt^+ strain is the donor of the cpDNA since the inheritance of resistance or sensitivity is dependent on the status of the mt^+ gene.

9. In a maternal effect, the *genotype* of the mother influences the *phenotype* of her immediate offspring in a non-Mendelian manner. The fact that all of the offspring (F_1) showed a dextral coiling pattern indicates that one of the parents (maternal parent) contains the *D* allele. Taking these offspring and seeing that their progeny (call these F_2) occur in a 1:1 ratio indicates that half of the offspring (F_1) are *dd*. In order to have these results, one of the original parents must have been *Dd* while the other must have been *dd*.

Parents: *Dd* X *dd*

Offspring (F_1): 1/2 *Dd*, 1/2 *dd*

(all dextral because of the maternal genotype)

Progeny (F_2):

All those from *Dd* parents will be dextral while all those from *dd* parents will be sinistral.

10. It appears as if some factor normally provided by the gs^+ allele is necessary for normal development and/or functioning of the female offspring's gonads. Without this product, the daughters are sterile, thus the term "grandchildless." Because the female provides so much vital material and information to the egg, including the cytoplasm necessary for germ line determination, it is not surprising that such maternal effect genes exist.

11. Since there is no evidence for segregation patterns typical of chromosomal genes and Mendelian traits, some form of extranuclear inheritance seems possible. If the *lethargic* gene is dominant then a maternal effect may be involved. In that case, when some of the F2 progeny would be hyperactive, because maternal effects are only temporary, affecting only the immediate progeny. If the lethargic condition is caused by some infective agent, then perhaps injection experiments could be used. If caused by a mitochondrial defect, then the condition would persist in all offspring of lethargic mothers, through more than one generation.

12. Developmental phenomena which occur early are more likely to be under maternal influence than those occurring late. Anterior/posterior and dorsal/ventral orientations are among the earliest to be established and in organisms where their study is experimentally and/or genetically approachable, they often show considerable maternal influence.

Chromosome Variations and Sex Determination

Vocabulary: Organization and Listing of Terms

Structures and Substances

Colchicine

Telomeres

rDNA

Processes/ Methods

Chromosome mutations (aberrations)

Sex determination in humans

- Y chromosome is male determining
- Klinefelter syndrome (XXY), 47 XXY
- Turner syndrome (XO), 45 X
- 47 XXX, 47 XYY, 48 XXXY, 48 XXYY,
  - 48 XXXX, 49 XXXXX,
  - 49XXXXY, 49 XXXYY,
  - 45 X/46 XY mosaic
- TDF (testis-determining factor)
- SRY (sex-determining region)
  - Sry (mice)
  - transgenic mice

Dosage compensation

- N-1 rule
- red-green color blindness
- anhidrotic ectodermal dysplasia

Sex determination in *Drosophila*

- nondisjunction
- triploid (3n)
- n + 1, n - 1, XXY, XO
- metafemale (superfemale)
- metamale
  - genic balance theory
    - 1X:2A (male)
    - 2X:2A, *etc.* (female)

Aneuploidy

- Monosomy (2n - 1)
  - haplo-VI (*Drosophila*)
  - partial monosomy
    - segmental deletions
    - Cri-du-Chat syndrome, 46,5p-
- Trisomy (2n + 1)
  - XXX (*Drosophila*, humans)
  - *Datura*
  - Down syndrome (G group)
    - trisomy 21 (47, 21+)
    - maternal age
    - amniocentesis
    - chorionic villus sampling (CVS)
      - familial Down syndrome
  - Patau syndrome
    - trisomy 13 (47, 13+)
  - Edwards syndrome
    - trisomy 18 (47, 18+)
- reduced viability
  - gametic level
  - embryonic level
  - spontaneously aborted fetuses

Euploidy (F6.1)

- monoploid (n)
- diploid (2n)
- polyploid
  - triploid (3n)
  - tetraploid (4n)
  - pentaploid (5n)
- autopolyploidy
  - autotriploids (3n)
    - complete nondisjunction
    - dispermic fertilization
    - tetraploid X diploid
  - autotetraploids
    - cold or heat shock
    - colchicine
- allopolyploidy
  - hybridization
  - allotetraploid (amphidiploid)
    - (cotton, *Triticale*)
  - protoplast fusion

Chromosome structure

- deletions
  - terminal, intercalary
  - loop (deficiency, compensation)
- duplications
  - gene redundancy
  - amplification

- rDNA
  - nucleolar organizer region (NOR)
  - micronucleoli in amphibians
- *Bar* eye in *Drosophila*
  - evolutionary aspects
    - Ohno
    - gene families
- rearrangements
  - inversions
    - "sticky ends"
    - paracentric
    - pericentric
    - heterozygotes
    - unusual pairing arrangments
      - inversion loops
      - dicentric chromatids
      - acentric chromatids
      - dicentric bridges
    - "suppression of crossing over"
  - translocations
    - reciprocal
    - unorthodox synapsis
    - semisterility
  - familial Down syndrome
    - frequency
    - Robertsonian translocation
    - 14/21 or D/G
  - fragile sites
    - X chromosome
    - *FMR-1*
      - trinucleotide repeat
    - Martin-Bell syndrome (MBS)

Concepts

- Significance
  - phenotypic variation
- Euploid, aneuploid
- Sex determination
  - dosage compensation
    - mosaicism
  - variation in chromosome number
  - genomic balance
    - plant vs. animal survival
  - sex chromosome balance
- Gene duplication (evolutionary aspects)
- Inversions
  - "suppression of crossing over"

F8.1. Illustration of the chromosomal configurations of duploid and aneuploid genomes of *Drosophila melanogaster.*

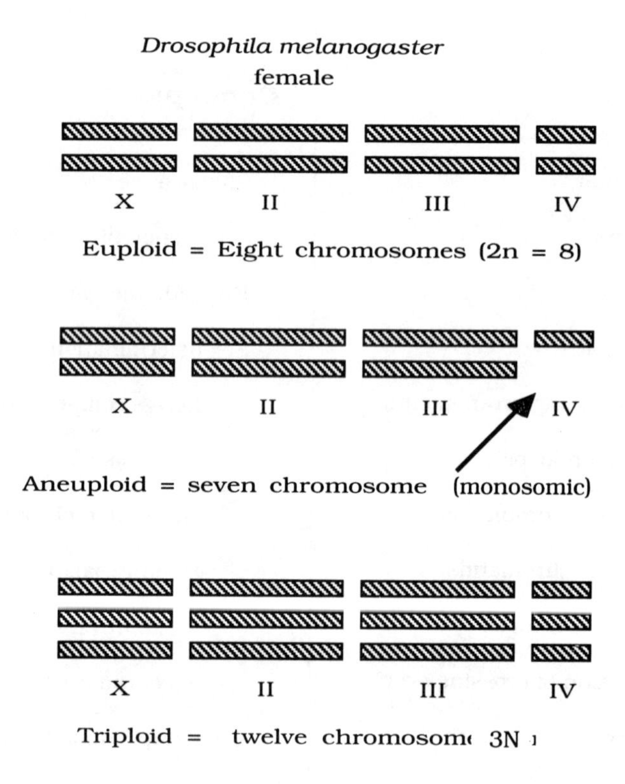

Solutions to Problems and Discussion Questions

1. Calvin Bridges (1916) studied nondisjunctional *Drosophila* that had a variety of sex chromosome complements. He noted that XO produced sterile males while XXY produced fertile females. In contrast, the Y chromosome was found to be male determining in humans. Individuals with the 47,XXY complement are males while 45,XO produces females.

In *Drosophila* it is the balance between the number of X chromosomes and the number of haploid sets of autosomes which determines sex. In humans there is a small region on the Y chromosome which determines maleness.

2. A *Barr body* is a darkly staining chromosome seen in some interphase nuclei of mammals with two X chromosomes. There will be one less Barr body than number of X chromosomes. The Barr body is an X chromosome which is considered to be genetically inactive. There is a simple formula for determining the number of Barr bodies in a given cell: N-1, where N is the number of X chromosomes.

Klinefelters syndrome (XXY)	= 1
Turners syndrome (XO)	= 0
47, XYY	= 0
47, XXX	= 2
48, XXXX	= 3

3. The *Lyon Hypothesis* states that the inactivation of the X chromosome occurs at random early in embryonic development. Such X chromosomes are in some way "marked" so that all progeny cells have the same X chromosome inactivated. Females will display mosaic retinas with patches of defective color perception. Under these conditions, their color vision may be influenced.

4. Normal development requires a complex and as yet poorly understood interplay of numerous gene products in time and space. Many gene products necessary for normal development are located on the X chromosome, which is normally present in two copies in females, one copy in males. If recessive lethal genes are present on the X chromosome, male conceptuses, perhaps having one lethal-bearing X chromosome, would not survive and female conceptuses which are XO (Turner syndrome) would also suffer the same fate. Another more complicated issue may be involved. Early in the development of a female, one of the normal two X chromosomes is randomly "shut off" so that each cell effectively has one X chromosome. Perhaps prior to the "shut off" process, female development actually requires that two relatively active X chromosomes are present. In the XO state, female embryos are not only vulnerable to recessive lethal genes on the X chromosome (as are males), but also suffer additional complications due to the absence of two X chromosomes very early in development.

5. Because synapsis of chromosomes in meiotic tissue is often accompanied by crossing over, it would be detrimental to sex-determining mechanisms to have sex-determining loci on the Y chromosome transferred, through crossing over, to the X chromosome.

6. With frequent exceptions, especially in plants, organisms typically inherit one chromosome complement (*haploid* = n = one representative from each homologous pair of chromosomes) from each parent. Such organisms are *diploid*, or 2n. When an organism contains complete multiples of the n complement (3n, 4n, 5n, etc.) it is said to be *euploid* in contrast to *aneuploid* in which complete haploid sets do not occur.

An example of an aneuploid is *trisomic* where a chromosome is added to the 2n complement. In humans, a trisomy 21 would be symbolized as 2n+1 or 47,21+.

Monosomy is an aneuploid condition in which one member of a chromosome pair is missing, thus producing the chromosomal formula of 2n-1. Haplo-IV is an example of monosomy in *Drosophila*. *Trisomy* is the chromosomal condition of 2n+1 where an extra chromosome is present.

Down syndrome is an example in humans (47, 21+).

Polyploidy refers to instances where there are more than two haploid sets of chromosomes in an individual cell. *Autopolyploidy* refers to cases of polyploidy where all the chromosomes in the individual originate from the same species. *Allopolyploidy* involves instances where the chromosomes originate from the hybridization of two different species, usually closely related.

Notice in the *Essentials* text the difference between *paracentric* and *pericentric* inversions. In paracentric inversions, the centromere is not included in the region bounded by the breakpoints, whereas in pericentric inversions, the breakpoints include the centromere.

7. The fact that there is a significant maternal age effect associated with Down syndrome indicates that nondisjunction in older females contributes disproportionately to the number of Down syndrome individuals. In addition, certain genetic and cytogenetic marker data indicate the influence of female nondisjunction.

8. As stated in the text, at least 20 percent of all conceptions are terminated in natural abortion. Of these, thirty percent show some chromosomal anomaly. Of the chromosomal anomalies that occur, approximately ninety percent are eliminated by spontaneous abortion. Aneuploidy contributes to the majority of spontaneous abortions.

Trisomy for every human chromosome has been observed, however, monosomy, the reciprocal meiotic event of trisomy, is rare. This observation probably results from gametic or early embryonic inviability.

9. For largely unknown reasons, plants tolerate increased levels of ploidy more than do animals. Polyploidy provides for greater phenotypic and genetic variation in plants, where concomitant reduction in fertility is overcome by various forms of asexual reproduction. Alloploidy provides an exceptional level of genetic mixing by allowing genomes from different species to interact in the formation of a new species. The paucity of polyploidy in animals may result from unknown developmental consequences mentioned above as well as upsets in mechanisms of sex determination.

10. Temperature shock applied during meiosis or colchicine applied during mitosis may lead to chromosome doubling. Colchicine interferes with spindle fiber formation.

11. American cultivated cotton has 26 pairs of chromosomes; 13 large, 13 small. Old world cotton has 13 pairs of large chromosomes and American wild cotton has 13 pairs of small chromosomes. It is likely that an interspecific hybridization occurred followed by chromosome doubling. These events probably produced a fertile amphidiploid (allotetraploid). Experiments have been conducted to reconstruct the origin of American cultivated cotton.

12. If the diploid chromosome number is 18, 2n=18, then in the somatic nuclei of

haploid individuals n = 9,

triploid (3n) = 27,

tetraploid (4n) = 36.

A trisomic has one extra chromosome, therefore it is 2n+1 = 19, and a monosomic is 2n-1 = 17. Refer to F6.1.

13. Examine various figures in the *Essentials* text and note the models for the formation of both terminal and intercalary deletions. Such deletions may also be caused by oblique synapsis and crossing over. Duplications may also be caused by this process as shown. In addition, an error in replication, perhaps by some "rocking motion" of a polymerase, may cause duplication.

Inversions may form in the process of loop formation and subsequent double breaks and healing (rejoining). Perhaps a crossover-like event occurs in the place of overlap. Translocations may originate by the proximity of non-homologous chromosomes and subsequent breakage and rejoining.

14. Basically the synaptic configurations produced by chromosomes bearing a deletion or duplication (on one homologue) are very similar. There will be point-for-point pairing in all sections which are capable of pairing. The section which has no homologue will "loop out" as shown in the *Essentials* text. Inversion loops, formed from the combination of an "outside" loop synapsing with an "inside" loop, are illustrated in the text.

15. Considering that there are at least three map units between each of the loci, and that only four phenotypes are observed, it is likely that genes *a b c d* are included in an inversion and crossovers which do occur among these genes are not recovered because of their genetically unbalanced nature. In a sense, the minimum distance between loci *d* and *e* can be estimated as 10 map units

(48 + 52/1000);

however this is actually the distance from the *e* locus to the breakpoint which includes the inversion. The "map" is therefore as drawn below:

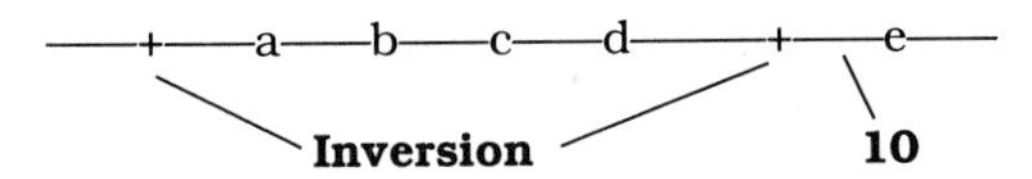

16. In a work entitled *Evolution by Gene Duplication*, Ohno suggests that gene duplication has been essential in the origin of new genes. If gene products serve essential functions, mutation and therefore evolution, would not be possible unless these gene products could be compensated for by products of duplicated, normal genes.

The duplicated genes, or the original genes themselves, would be able to undergo mutational "experimentation" without necessarily threatening the survival of the organism.

17. A Turner syndrome female has the sex chromosome composition of XO. If the father had hemophilia it is likely that the Turner syndrome individual inherited the X chromosome from the father and no sex chromosome from the mother. If nondisjunction occurred in the mother, either during meiosis I or meiosis II, an egg with no X chromosome can be the result.

18. The primrose, *Primula kewensis*, with its 36 chromosomes, is likely to have formed from the hybridization and subsequent chromosome doubling of a cross between the two other species, each with 18 chromosomes.

19. In the trisomic, segregation will be "2 X 1" as illustrated below:

P_1:

b/b/b X *B/B*

gametes: (*bb*) (*b*) (*B*)

F_1:

$B/b^1/b^2$ X B/b
(normal bristles) (normal bristles)

Notice that there are several segregation patterns created by the trivalent at anaphase I.

gametes:

b^+	b^+
b^1b^2	b
b^+b^1	
b^+b^2	
b^2	
b^1	

F_2:

b^+b^+	= normal bristles
$b^+b^1b^2$	= normal bristles
$b^+b^+b^1$	= normal bristles
$b^+b^+b^2$	= normal bristles
b^+b^2	= normal bristles
b^+b^1	= normal bristles
b^+b	= normal bristles
bb^1b^2	= bent bristles
b^+bb^1	= normal bristles
b^+bb^2	= normal bristles
bb^2	= bent bristles
bb^1	= bent bristles

20. Refer to the *Essentials* text and notice that the phenotypic mosaicism is dependent on the heterozygous condition of genes on the two X chromosomes. Dosage compensation and the formation of Barr bodies occurs only when there are two or more X chromosomes. Males normally have only one X chromosome, therefore such mosaicism can not occur. Females normally have two X chromosomes. There are cases of male calico cats which are XXY.

21. **(a)** In all probability, crossing over in the inversion loop of an inversion (in the heterozygous state) had produced defective, unbalanced chromatids, thus leading to stillbirths and/or malformed children.

(b) It is probable that a significant proportion (perhaps 50%) of the children of the man will be similarly influenced by the inversion.

(c) Since the karyotypic abnormality is observable, it may be possible to detect some of the abnormal chromosomes of the fetus by amniocentesis or CVS. However, depending on the type of inversion and the ability to detect minor changes in banding patterns, all abnormal chromosomes may not be detected.

22. Since there is a region of synapsis close to the SRY-containing section on the Y chromosome, crossing over in this region would generate XY translocations which would lead to the condition described.

23. Because the Klinefelter son is $X^{g1}X^{g2}Y$, he must have obtained the X^{g1} allele and the Y chromosome from the father. Thus non-disjunction must have occurred during meiosis I in the father.

24. **(a)** The father must have contributed the abnormal X-linked gene.

(b) Since the son is XXY and heterozygous for anhidrotic dysplasia, he must have received both the defective gene and the Y chromosome from his father. Thus non-disjunction must have occurred during meiosis I.

(c) Ths son's mosaic phenotype is caused by X-chromosome inactivation, a form of dosage compensation in mammals.

Sample Test Questions (with detailed explanations of answers)

The purpose of these Sample Test Questions is to present a slightly different style of question. Set aside several hours of study time, perhaps a week before the first examination. Attempt to work these questions, five or so per hour, under test conditions. Write down your answers on paper, then, *after* you have finished the "test," check your answers. If you are having difficulty, then you are weak in the concept areas listed for each question. If you have made mistakes, take comfort, there are many places to make mistakes on these problems. Some of the students who made the same mistakes are now practicing geneticists! Assuming that the first (or first and second) examination covers Chapters 1 through 8, these questions should apply.

Question 1. The mosquito, *Culex pipiens*, has a diploid chromosome number of 6. Assume that one chromosome pair is metacentric, and the other two pairs are acrocentric.

(a) Draw chromosomal configurations which one would expect to see at the following stages: primary oocyte (metaphase I), secondary spermatocyte (metaphase II).

(b) Assuming that a G_1 nucleus in *Culex* contains about 20 picograms (pg) of DNA, how much DNA would you expect in the following nuclei; Primary Spermatocyte, First Polar Body, Secondary Oocyte, Ootid in G_1 phase.

(c) Assume that a female mosquito is heterozygous for the recessive gene *wavy bristles* (symbolized as *wb*) and this gene locus is on an acrocentric chromosome. Draw an expected mitotic metaphase with the appropriate genetic labeling pattern.

Concepts:

chromosome mechanics

mitosis, meiosis

symbolism

DNA content (cell cycles)

Answer 1. This question is intended to determine your understanding of mitosis, chromosome morphology, symbolism, the positioning of genes on chromosomes, and the changes in DNA content through the cell cycles.

(a) Since the diploid chromosome number is six, there will be three bivalents, one involving metacentrics, and two involving acrocentrics in a primary oocyte. We can draw the chromosomes of the primary oocyte as follows:

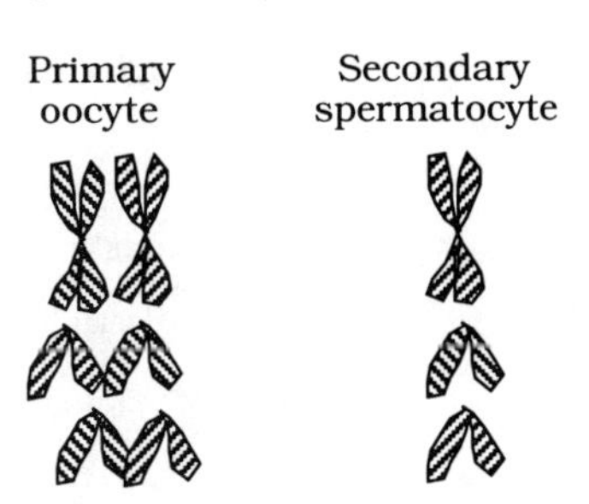

Note: homologous chromosomes will have "slashes" going in opposite directions because of the shadowing used to depict maternal and paternal chromosomes.

Since secondary spermatocytes arise after meiosis I, there should be only dyads, one metacentric and two acrocentric, and they should be aligned end-to-end as indicated in the drawing above.

(b) Given that there are about 20 picograms of DNA in a G_1 nucleus, we would expect there to be 40pg in a G_2 nucleus (after S phase) and 40pg to the point where homologous chromosomes separate in meiosis I. Secondary spermatocytes and secondary oocytes (as well as first polar bodies) should therefore, each have 20pg of DNA. After meiosis II the resulting nuclei should have 10pg each. If you understand events at interphase and in meiosis, this question is easy to answer. Carefully examine the figure below to understand events during the interphase as far as DNA content is concerned. Then examine F2.3 to see how chromosomes are behaving in meiosis. From this information you should see the answers as follows:

Primary spermatocyte	= 40 pg
First Polar Body	= 20pg
Secondary oocyte	= 20pg
Ootid (in G_1)	= 10pg

It might be helpful to view changes in DNA content in graphic form:

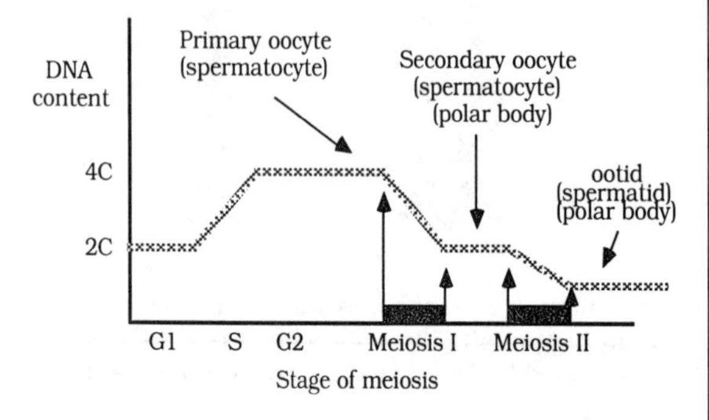

The "C" stands for "complements" of DNA.

(c) If the mosquito is heterozygous for the recessive gene *wavy bristles* (*wb*) then it would have the genotype *Wb/wb*. Because there are four letters here representing the two genes, the slash between the symbols helps us to understand that there are only two genes being discussed.

We are asked to draw an acrocentric, mitotic metaphase chromosome complement in this heterozygous insect. We are expected to place the gene symbols on the chromosomes. Recall that there is no synapsis of homologous chromosomes in mitotic cells, therefore, the chromosomes should not be placed side-by-side. Since sister chromatids are *identical* and homologous chromosomes are *similar*, we should draw the figure as follows:

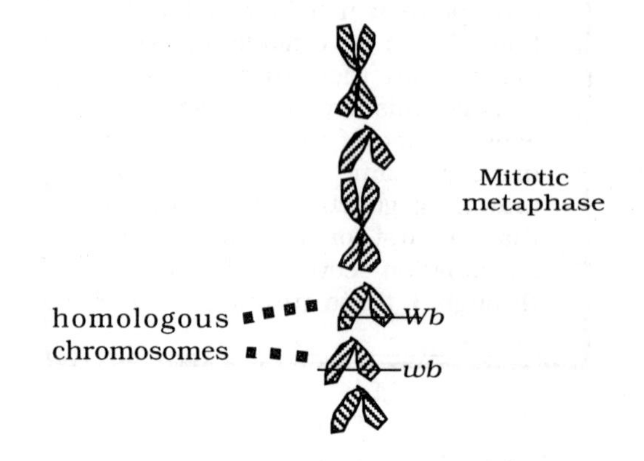

Common errors:

- **incorrect number of chromosomes**
- **incorrect chromosome morphology**
  - **metacentric, acrocentric**
- **poor relationship of DNA content to cells**
- **inappropriate symbols**
- **inappropriate placement of genes**

Question 2. In humans, chromosome #1 is large and metacentric, the X chromosome is medium in size and submetacentric (submedian), while the Y chromosome is small and acrocentric. Assume that you were microscopically examining human chromosomes at the stages given below.

(a) Illustrate (draw) the above-mentioned (#1, X and/or Y) chromosomes and/or pairs at the stages given (several different configurations may be applicable in some cases):

Metaphase I (Primary Oocyte):

First Polar Body:

Secondary Spermatocyte:

Secondary Oocyte:

(b) The Rh blood group locus is on Chromosome #1. Individuals are *DD* or *Dd* if Rh^+ and *dd* if Rh^-. The locus for glucose-6-phosphate-dehydrogenase deficiency (G6PD) is located on the X chromosome. There are two alternatives at this locus, + and -. For each of the above cells, place genes (using symbolism given) on chromosomes if the female is heterozygous at both the Rh and G6PD loci. Do the same for the secondary spermatocyte (above) assuming that the male is Rh-, and + for the G6PD locus.

(c) Assume that the average DNA content per G_1 nucleus in humans is 6.5 picograms. For the nuclei (including the entire chromosome complement for each nucleus) presented, give the expected DNA content:

Metaphase I (Primary Oocyte) :

First Polar Body:

Secondary Spermatocyte:

Secondary Oocyte:

Concepts:

chromosome mechanics

mitosis, meiosis

symbolism

DNA content (cell cycles)

Answer 2. (a,b) Recall that a metacentric chromosome has "arms" of approximately equal length, while submetacentric and acrocentric chromosomes have arms of unequal length.

Metaphase I (Primary Oocyte): homologous chromosomes are replicated and synapsed. There will be two X chromosomes present because oocytes occur in females. On the metacentric chromosomes (#1), place, such that sister chromatids are identical, the *Dd*. Place the + - alternatives on the X chromosomes.

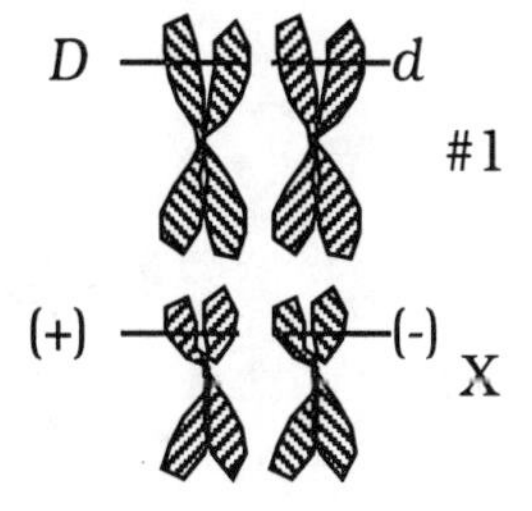

First Polar Body: the first polar body is a product of meiosis I, after homologous chromosomes have migrated to opposite poles. At this stage, dyads are present. Because females produce polar bodies, there should be an X chromosome present. Because the female is heterozygous, there are several possible answers. Note that there is only one representative of each allele for each gene pair.

One possibility

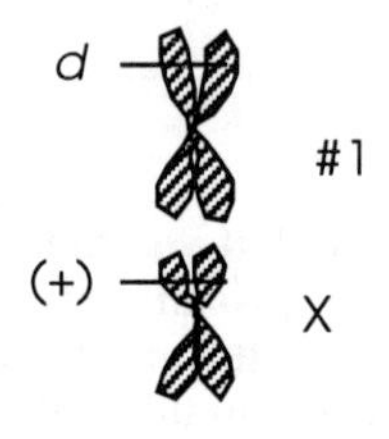

Secondary Spermatocyte: a secondary spermatocyte will have the same chromosome configuration as a First Polar Body. Both are products of meiosis I and dyads should be present. Because spermatocytes occur in males, there will either be an X chromosome or a Y chromosome present. There are therefore two possible answers. Regarding the genetic constitution of these cells, as stated in the problem, we are to assume that the male is Rh^- and + for the G6PD locus. Since this locus is on the X chromosome, only one genotype (regarding the X chromosome) can be presented.

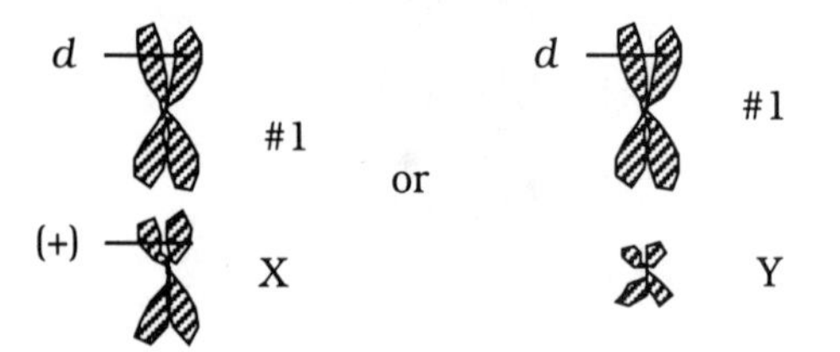

Secondary Oocyte: being a product of meiosis I, dyads will be present. Because oocytes occur in females, there should be an X (not a Y) chromosome present. The genetic labeling pattern for the secondary oocyte will be the same as for the first polar body.

One possibility

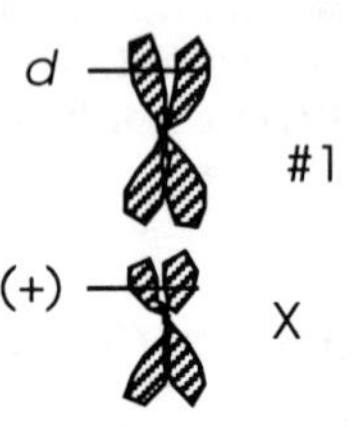

(c) Referring to the figure for part (b) in the above problem, if a G_1 nucleus contains 6.5pg DNA, then the following DNA contents are expected.

Metaphase I (Primary Oocyte):	13pg
First Polar Body:	6.5pg
Secondary Spermatocyte:	6.5pg
Secondary Oocyte:	6.5pg

Common errors:

incorrect number of chromosomes

incorrect chromosome morphology

metacentric, acrocentric

poor relationship of DNA content to cells

inappropriate symbols

inappropriate placement of genes

Question 3. Assume that you are examining a cell under a microscope and you observe the following as the total chromosomal constituents of a nucleus. You know that 2n=2 in this organism, that all chromosomes are telocentric, and that each G_1 cell nucleus contains 8 picograms of DNA.

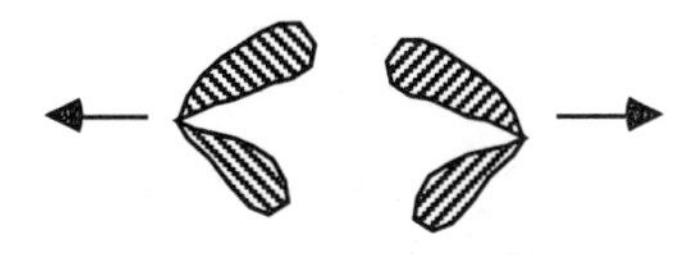

(a) Circle the correct stage for this cell:

anaphase of mitosis,

anaphase of meiosis I,

anaphase of meiosis II,

telophase of mitosis.

(b) How many picograms of chromosomal DNA would you expect in the cell shown above?

Concepts:

chromosome morphology

telocentric, etc.

anaphase configurations

chromosome mechanics

meiosis, mitosis

DNA content in cell cycles

Answer 3.

(a) Since the cell contains only two chromosomes (2n=2) and there are two chromosomes pictured, it can not represent a cell in the second phase (II) of meiosis. The chromosomes are telocentric which means that the centromere is at the end of the chromosome. When pulled at anaphase, two sideways "**V**s" or "**< >**" would be expected and the cell would be at the anaphase stage of meiosis I.

The only other possibility to produce the "**< >**" figure would be a metaphase chromosome at anaphase of mitosis or anaphase II of meiosis. However, these possibilities are negated because the chromosomes are stated as being telocentric.

(b) In order to get the correct answer for the second part one must consider that, because of the S-phase, at anaphase I the DNA complement is twice that of a G_1 cell. Therefore the correct answer is 16pg DNA.

Common errors:

confusion on:

significance of *telocentric*

significance of chromosome number

many students consider the chromosomes to be metacentric

Question 4. The genes for *singed bristles* (*sn*) and *miniature wings* (*m*) are recessive and located on the X chromosome in *Drosophila melanogaster.* In a cross between a singed-bristled, miniature-winged female and a wild type male, all of the male offspring were singed-miniature.

(a) Draw meiotic metaphase I chromosomal configurations which represent the X and/or Y chromosomes of the parental (singed-miniature female and wild type male) flies. Place gene symbols (*sn*, *m*) and their wild type alleles (sn^+, m^+) on appropriate chromosomes.

(b) Draw a mitotic metaphase chromosomal configuration that represents the X and/or Y chromosomes of the F1 male. Place gene symbols (*sn*, *m*) on appropriate chromosomes.

(c) Most of the female offspring from the above-mentioned cross were phenotypically wild type; however one exceptional female was recovered which had singed bristles and miniature wings. Given that meiotic nondisjunction accounted for this exceptional female, would you expect it to have occurred in the parental male of parental female?

(d) Draw a meiotic, labeled (with gene symbols) circumstance and division product(s) which could account for the exceptional female described above. (Confine your drawing to X chromosomes only).

Concepts:

chromosome mechanics

meiosis

symbolism

DNA content (cell cycles)

meiotic nondisjunction

sex determination

Answer 4.

(a) The female parent would have the following labeled chromosomal symbolism remembering that at meiotic metaphase I, chromosomes are doubled, condensed, and synapsed.

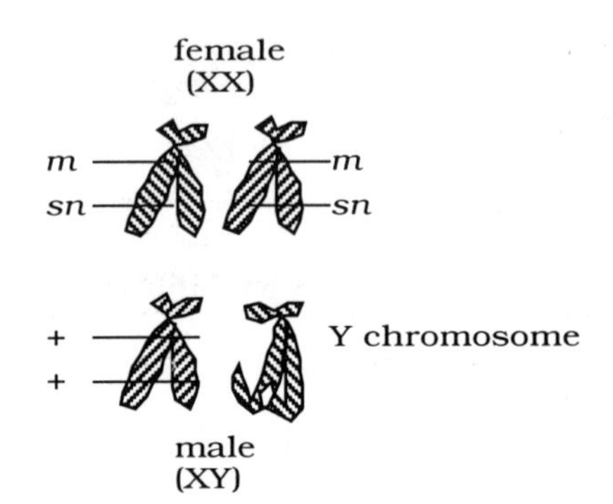

(b) In general, mitotic metaphase chromosomes are not synapsed, although in *Drosophila* mitotic chromosomes do pair. To avoid confusion and to be consistent with what is expected in other organisms, the mitotic chromosomes will not be drawn in the paired state.

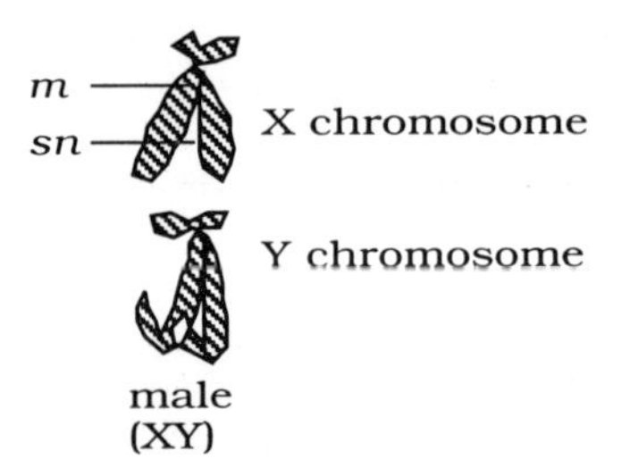

(c) All of the female offspring from the above cross should be heterozygous and phenotypically wild type. The one exceptional female could have resulted from maternal nondisjunction at meiosis I or II, thus producing an egg cell with two X chromosomes, each containing the *sn* and *m* genes. When fertilized by a sperm cell carrying the Y chromosome (along with the normal haploid set of autosomes) an $X^{sn\ m}X^{sn\ m}Y$ female is produced.

(d)

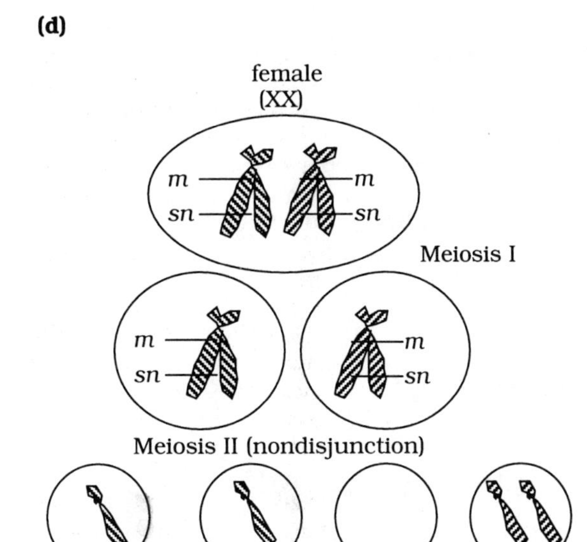

Common errors:

incorrect chromosome morphology

X, Y chromosmes

inappropriate symbols

inappropriate placement of genes

problems with meiotic nondisjunction

Question 5. The red fox (*Vulpes vulpes*) has 17 pairs of somewhat long chromosomes. The Arctic fox (*Alopex lagopus*) has 26 pairs of somewhat shorter chromosomes.

(a) If a female red fox is crossed with a male Arctic fox, what will be the chromosome number in the somatic tissues of the hybrid?

(b) Assume that a somatic G_1 nucleus of the Arctic fox contains 12 picograms of DNA while a somatic G_1 nucleus of the red fox contains 8 picograms of DNA. How much nuclear DNA could you expect in a G_2 somatic nucleus of the hybrid?

Concepts:

meiosis and chromosome numbers

DNA content

Answer 5.

(a) Since the chromosome numbers are given in *pairs*, recall that during meiosis each gamete contains one chromosome of each pair. If the red fox has 17 pairs of chromosomes, then each gamete will contain 17 chromosomes. For the Arctic fox, each gamete should contain 26 chromosomes. A zygote is produced from the union of the parental gametes, therefore it should contain 43 chromosomes (17 + 26). It turns out that some such hybrids are viable but usually sterile because of developmental and chromosomal alignment and segregational problems at meiosis.

(b) If G_1 nuclei contain 12pg and 8pg DNA, then the gametes produced from these organisms will contain 6 and 4pg DNA respectively. Combining these gametes gives 10pg for a G_1 cell. For a G_2 cell there should be 20pg DNA.

Common errors:

confusion with *pairs* of chromosomes

gametic chromosome number

confusion with uneven number of chromosomes

confusion with *somatic* cells

Question 6. Red-green colorblindness is inherited in man as an X-linked, recessive gene. Using the symbols below, draw a pedigree which is consistent with the following statements.

A phenotypically normal woman is married to a phenotypically normal man. The woman's parents are phenotypically normal but her maternal grandfather is colorblind. The woman's paternal grandparents as well as her maternal grandmother are phenotypically normal.

male	=	□
female	=	○
Rg	=	**normal color sight**
rg	=	**colorblind**

What is the probability that the first son born to the woman will be phenotypically normal (not be colorblind)?

Concepts:

sex-linked inheritance (X-linked)

pedigree construction

probability (product rule)

Answer 6.

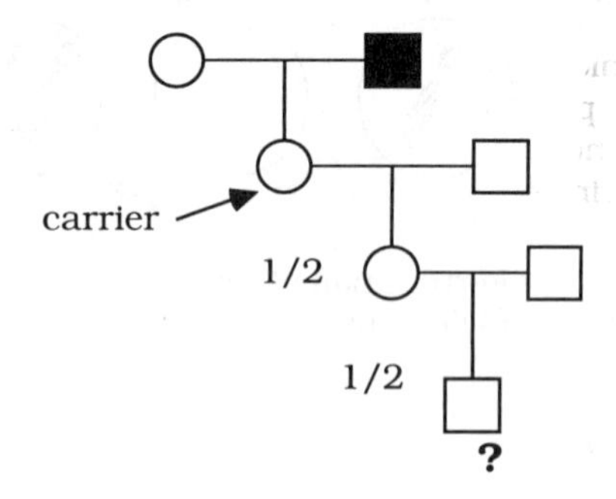

Because the maternal grandfather was colorblind, the woman's mother is a carrier for this X-linked gene ($X^{Rg}X^{rg}$). The woman therefore has a 1/2 chance of inheriting the X^{rg} chromosome from her mother and a 1/2 chance of passing this X^{rg} chromosome to her son. The chance that the son will receive the X^{rg} chromosome is therefore 1/4 (1/2 X 1/2). However, the question asks for the probability that the son will be normal.

The answer is therefore 1 minus 1/4 which equals 3/4.

Common errors:

difficulty in setting up pedigree

inability to see independent probabilities

multiplication of independent probabilities

seeing that the *normal* is requested

Question 7. In a *Drosophila* experiment a cross is made between a homozygous wild type female and a tan-bodied (mutant) male. All the resulting F_1 flies were phenotypically wild type. Adult flies of the F_2 generation (from a mating of the F_1's) had the following characteristics:

Sex	*Phenotype*	*Number*
Male	wild	346
Male	tan	329
Female	wild	702

(a) Using conventional symbolism, illustrate the genotype, *on an appropriate chromosomal configuration*, of a secondary oocyte nucleus of one of the F_1 females. Be certain to distinguish the X chromosomes from the autosomes. Note: *Drosophila melanogaster* has a diploid chromosome number of 8.

(b) Using the same conventional symbolism, give the genotype *on an appropriate chromosomal configuration* of a primary spermatocyte of the tan-bodied F_2 males.

Concepts:

sex-linked inheritance (X-linked)

chromosome mechanics

meiosis

conventional symbolism

Answer 7.

(a) First one must determine whether the gene for *tan body* is X-linked or autosomal (not on the sex chromosome). Because half of the F_2 males are mutant and half are wild type, and all the females are wild, the gene for *tan body* is behaving as X-linked. The F_1 female is heterozygous, therefore she should have either of the alleles (t, t^+) on the one X chromosome (a secondary oocyte has one representative of each chromosomal pair) in the following arrangement. Because *Drosophila* has 8 chromosomes, each secondary oocyte should have four chromosomes (including the X chromosome).

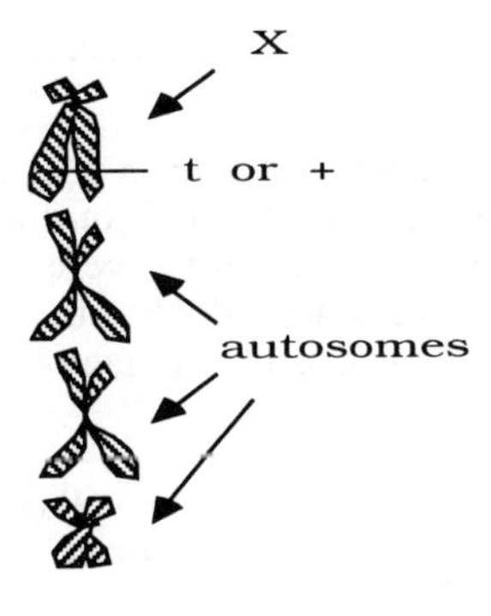

(b) A primary spermatocyte has the chromosomes in a doubled, condensed, and synapsed state. A tan-bodied male should have an X (containing a *t* gene) and Y chromosome as well as a diploid complement of autosomes.

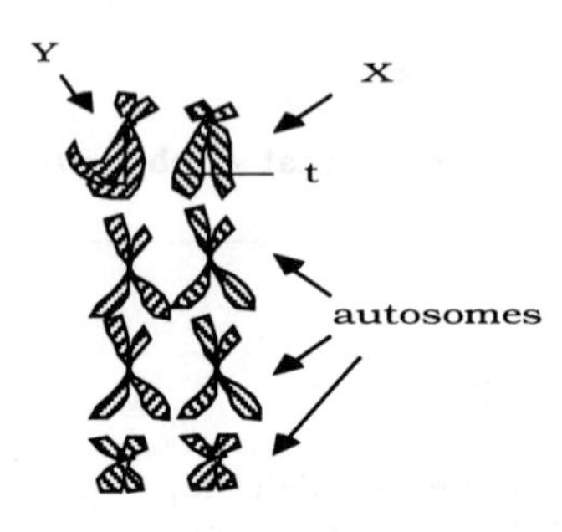

> **Common errors:**
>
> **difficulty in recognizing X-linked inheritance**
>
> **problems with placing genes on chromosomes**
>
> **problems with visualizing genome**

Question 8. *Gray* seed color in peas is dominant to *white*. Assume that Mendel conducted a series of experiments where plants were crossed and offspring classified according to the table below. What are the most probable genotypes of each parent?

Parents			*Progeny*	
			gray	*white*
(a) gray	X	white	81	79
(b) gray	X	gray	120	42
(c) white	X	white	0	50
(d) gray	X	white	74	0

> **Concepts:**
>
> **Mendelian genetics**
>
> **monohybrid cross**
>
> **dominance/recessiveness**
>
> **3:1 and 1:1 ratios**

Answer 8. First, assign gene symbols:

G = gray, *gg* = white

(a) Since there is an approximate 1:1 ratio in the progeny, the parental genotypes are *Gg* X *gg*.

(b) A 3:1 ratio is apparent, therefore the parental genotypes are *Gg* X *Gg*.

(c) Because there are no gray phenotypes and *gray* is the dominant allele, the parental genotypes must be *gg* X *gg*.

(d) Since there are no white types and the sample is sufficiently large, it is very likely that the parental genotypes are *GG* X *gg*.

> **Common errors:**
>
> **students usually have only minor problems with this type of question**
>
> **some careless, random mistakes**

Question 9. Hemophilia (type A) is recessive and X-linked in humans whereas the ABO blood groups locus is autosomal. Assume that the following matings were examined for the transmission of these genes. Give the expected phenotypes and numbers assuming 800 offspring are produced.

Group A: Females heterozygous for hemophilia with blood type AB mated to normal males with blood group O.

Group B: Females heterozygous for hemophilia with blood type AB mated to males with hemophilia and blood group AB.

Concepts:

- **sex-linkage (X-linked)**
- **autosomal inheritance**
- **dihybrid situation**
- **incomplete dominance**
- **complete dominance**

Answer 9. Set up the crosses with an appropriate symbol set such as the following:

h = hemophilia

H = normal allele

I^AI^B = AB blood group

I^oI^o = O blood group.

Group A.

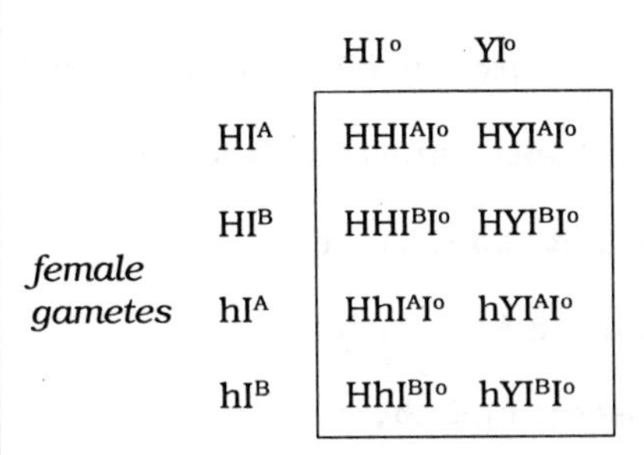

		male gametes	
		HI^o	YI^o
	HI^A	HHI^AI^o	HYI^AI^o
	HI^B	HHI^BI^o	HYI^BI^o
female gametes	hI^A	HhI^AI^o	hYI^AI^o
	hI^B	HhI^BI^o	hYI^BI^o

Collecting phenotypes gives:

1/4 female, normal, A blood (200)

1/4 female, normal, B blood (200)

1/8 male, normal, A blood (100)

1/8 male, normal, B blood (100)

1/8 male, hemophilia, A blood (100)

1/8 male, hemophilia, B blood (100)

Group B. In this example the forked-line method will be used. Consider what will be happening for the *hemophilia* locus independently from the blood group locus.

1/4 females, normal — 1/4 A blood (50)
 2/4 AB blood (100)
 1/4 B blood (50)

1/4 females, hemophilia – 1/4 A blood (50)
 2/4 AB blood (100)
 1/4 B blood (50)

1/4 males, normal — 1/4 A blood (50)
 2/4 AB blood (100)
 1/4 B blood (50)

1/4 males, hemophilia — 1/4 A blood (50)
 2/4 AB blood (100)
 1/4 B blood (50)

Common errors:

difficulty with any dihybrid situation

X-linked with autosomal inheritance

incomplete dominance

calculating frequencies

observed numbers

Question 10. For the cross *PpRr X pprr* where complete dominance and independent assortment hold, assume that you received the following results and you wished to determine whether they differ significantly (in a statistical sense) from expectation.

PR phenotypes = 40

Pr phenotypes = 10

pR phenotypes = 20

pr phenotypes = 30

(a) State the null hypothesis associated with this test of significance.

(b) How many degrees of freedom would be associated with this test of significance?

(c) Assuming that a Chi-Square value of 20.00 is arrived at in this test of significance, do you accept or reject the null hypothesis?

Degrees of Freedom	*P = 0.05*
1	3.84
2	5.99
3	7.82
4	9.49
5	11.07

Concepts:

Mendelian patterns

1:1:1:1 ratio

null hypothesis

expected values

χ^2 analysis

interpretation of χ^2

Answer 10.

(a) An appropriate null hypothesis for this example would be that the observed (measured) values do no not differ significantly from the predicted ratio of a 1:1:1:1. One might also say that any deviation between the observed and predicted values is due to chance and chance alone.

(b) Because there are four classes being compared, there will be three degrees of freedom.

(c) Given that the Chi-square value of 20.00 is considerably greater than 7.82 (for three degrees of freedom) the null hypothesis should be rejected and the conclusion should be that the observed values differ significantly from the predicted values based on a 1:1:1:1 ratio.

Common errors:

inability to see a 1:1:1:1 ratio

development of the expected ratios

interpreting probability values from table

Question 11. Explain the processes, genotypic, chromosomal, and developmental, which would lead to a bilateral gynandromorph in *Drosophila melanogaster* in which the male half of the fly has white eyes and singed bristles, while the female half is phenotypically wild type.

Concepts:

- **sex-linked inheritance (X-linked)**
- **gynandromorph production**
- **sex determination in *Drosophila***
- **insect development**
- **mitosis, nondisjunction**

Answer 11. In order for this type of fly to occur, the zygote must start out as a heterozygote in which both mutant genes are on one homologue and wild type alleles are on the other. In addition, one of the wild type chromosomes must get "lost" at the first mitotic division, thus making the female half $X^{++}X^{w\,sn}$ and the other half $X^{w\,sn}$ O. The diagram below explains this situation.

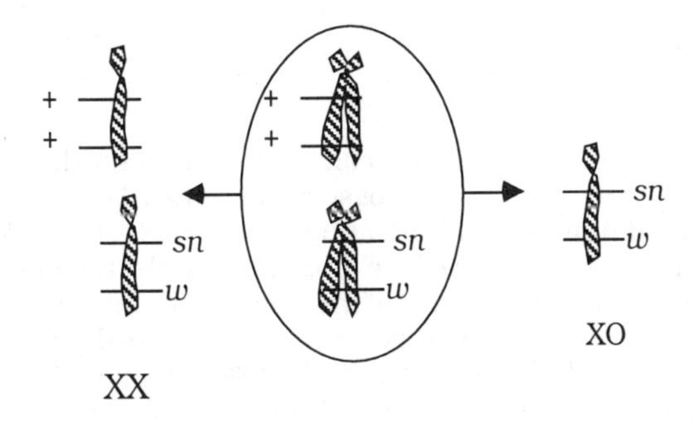

Because XX nuclei produce female tissue and XO nuclei produce male tissue, the phenotypes of the two sides are thus described. Developmentally, once the cleavage nuclei reach the peripheral areas of the egg to form a blastoderm, they become committed as to their adult fate. Because there is little "wandering" of nuclei either during their migration to the egg periphery or after they reach the periphery, the male/female boundary is quite clean.

Common errors:

- **difficulty in setting up the problem**
- **dealing with mitotic nondisjunction**
- **embryonic development of *Drosophila***

Question 12. Describe similarities and differences between *discontinuous* and *continuous* traits at the *molecular* and *transmission* levels.

Concepts:

- **interaction of gene products**
- **relationship between genotype and phenotype**
- **multi-factor inheritance**

Answer 12. At the *molecular level* one could consider that in discontinuous inheritance the gene products are acting fairly independently of each other, thereby providing a 9:3:3:1 ratio in a dihybrid cross for example. Genotypic classes

A-B-, *A-bb*, *aaB-*, and *aabb*

can be clearly distinguished from each other because the gene products from the *A* locus produce distinct influences on the phenotypes as compared to those gene products from the *B* locus. Exceptions exist where epistasis and other forms of gene interaction occur. In discontinuous inheritance one would consider each locus as providing a *qualitatively* different impact on the phenotype. For instance, even though the *brown* and *scarlet* loci interact in the production of eye pigments in *Drosophila*, each locus is providing qualitatively different input.

In continuous inheritance, we would consider each involved locus as having a quantitative input on the production of a single characteristic of the phenotype. In addition, although it may not always be the case, we would consider each gene product as being qualitatively similar. Under this model, the *quantity* of a particular set of gene products, influenced by a number of gene loci, determines the phenotypic characteristic.

At the *transmission level* one sees "step-wise" distributions in discontinuous inheritance but "smoother" or more bell-shaped distributions in continuous inheritance as shown in the figure below. For instance, in a dihybrid situation (*AaBb* X *AaBb*) where independent assortment holds, one would obtain a 9:3:3:1 ratio (assuming no epistasis, etc.) under a discontinuous mode but a 1:4:6:4:1 where genes (or gene products) are acting additively (continuous inheritance). Both patterns are formed from normal Mendelian principles of segregation, independent assortment, and random union of gametes. It is the manner in which the genes (or gene products) interact which distinguishes discontinuous from continuous inheritance.

Common errors:

difficulty with "molecular level" of the question

confusion over differences and similarities relating discontinuous and continuous patterns

Question 13. Assume that there are 18 map units between two loci in the mouse and that you are able to microscopically observe meiotic chromosomes in this organism. If you examined 150 primary oocytes, in how many would you expect to see a chiasma between the two loci mentioned above?

Concepts:

crossing over mechanisms

meiosis

gene mapping

chromosome mechanics

Answer 13. The basis of the solution is to recall that crossing over occurs at the "four-strand stage" (after the S-phase) and each chiasma involves only two of the four chromatids present in each tetrad. Therefore for each chiasma only two of the four, or 1/2, of the chromatids are crossover chromatids.

$$\frac{\text{Number of crossover progeny}}{\text{Total number of progeny}} \times 100$$

Gene mapping basically is the process of dividing the number of crossover chromatids by the total number of chromatids. Since each chiasma involves only two of the four chromatids, the map distance must be half of the chiasma frequency. If there are 18 map units between two genes then the chiasma frequency would be 36%. If one examined 150 primary oocytes one would therefore expect to see 0.36 X 150, or 54 cells with a chiasma between the two loci.

Common errors:

problem seeing relationships

chiasma, map units

problems "seeing" meiosis

visualization of crossing over

Question 14. Given below are four dihybrid crosses between various strains of *Drosophila*. To the right of each are map distances known to exist between the genes involved. For each cross give the phenotypes of the offspring and the percentages expected for each.

	Mating		
	Female	Male	Map distance
(a)	*AB/ab*	*ab/ab*	20
(b)	*Pq/pQ*	*pq/pq*	50
(c)	*DB/db*	*db/db*	0
(d)	*ab/ab*	*AB/ab*	20

Concepts:

linkage and crossing over

computation of map units

complete linkage

independent assortment

lack of crossing over in males

Answer 14. The key to solving these type of "reverse mapping" problems is to keep in mind that a map unit is computed by the equation

$$\frac{\text{Number of crossover progeny}}{\text{Total number of progeny}} \times 100$$

and that if the map distance is given it is easy to determine the percentages of parental and crossover offspring. Remember that there are two classes of crossovers and two classes of parentals from each cross.

(a)

AB/ab = 40%, *ab/ab* = 40% (parentals)

Ab/ab = 10%, *aB/ab* = 10% (crossovers)

(b)

Pq/pq = 25%, *pQ/pq* = 25% (parentals)

PQ/pq = 25%, *pq/pq* = 25% (crossovers)

Notice that this is independent assortment.

(c) *DB/db* = 50%, *db/db* = 50% (all parentals)

(d) *AB/ab* = 50%, *ab/ab* = 50% (all parentals)

The reason that there are all parentals and no crossovers in this cross is that there is no crossing over in male *Drosophila*. With no crossing over, the *AB/ab* chromosomes in the male are passed to gametes without crossovers.

Common errors:

difficulty in going from map units to frequencies of classes of offspring

failure to see that there are two parental and two crossover classes

minor, careless mistakes

Question 15. Direction of shell coiling in the land snail, *Limnaea peregra*, is determined by alleles at a single locus: *dextral* (right) = *DD* or *Dd; dd* = *sinistral* (left). However, a maternal effect is present such that the genotype of the mother determines the direction of coiling (phenotype) of the immediate offspring. Given the following crosses, write the genotypes and phenotypes in the spaces provided. Be certain to indicate which genotypes go with which phenotypes.

	Source of sperm	***Source of egg***
Cross #1	***DD***	***dd***

Offspring genotype(s): *Offspring phenotype(s):*

Cross #2	***dd***	***Dd*** (sinistral)

Offspring genotype(s): *Offspring phenotype(s):*

Cross #3	***Dd***	***Dd***

Offspring genotype(s): *Offspring phenotype(s):*

Concepts

maternal effects

Answer 15: The definition provided in the initial problem can be applied directly to the solution of this problem. The *genotype* of the mother determines the direction of coiling (phenotype) of the immediate offspring. In this case, one merely assigns the genotypes on the basis or normal Mendelian principles, then the phenotypes based on the genotype of the mother.

Notice in Cross 2, the *Dd* has a sinistral phenotype. While this may confuse some students, remember that the phenotype is determined by the *maternal* genotype. When early developmental events are involved, often the maternal genotype will have a significant influence over those events because the mother makes the egg.

Cross #1:

Offspring genotype(s):	*Offspring phenotype(s):*
Dd	all sinistral

Cross #2:

Offspring genotype(s):	*Offspring phenotype(s):*
Dd *dd*	all dextral

Cross #3:

Offspring genotype(s):	*Offspring phenotype(s):*
DD *Dd* *dd*	all dextral

Common errors:

When students are reminded of the nature of a maternal effect, that is, given a definition, there are very few errors. However, when they are asked to work the problem without the definition given, many have difficulty. Cross #2 causes most of the problems.

Notes

Notes

Notes

Notes

DNA - The Physical Basis of Life

Vocabulary: Organization and Listing of Terms

Historical

Miescher (nuclein) - 1868
- genetic material
  - proteins
  - nucleic acids
    - tetranucleotide hypothesis
    - base ratios
      - Erwin Chargaff (1940s)
    - transforming principle
      - Griffith (1927)
      - Avery *et al.* (1944)
    - T2 bacteriophage (phage)
      - Hershey and Chase (1952)
        - ^{32}P, ^{35}S
    - Watson and Crick (1953)
      Franklin, Wilkins

Structures and Substances

Messenger RNA (mRNA)

Transfer RNA (tRNA)

Ribosomal RNA (rRNA)
- ribosome

Proteolytic enzymes

Ribonuclease

Deoxyribonuclease

Protoplasts (spheroplasts)

Insulin

Human ß-globin gene

RNA core

Coat protein

Qß RNA replicase

Reverse transcriptase

Nucleic acids
- nucleotide

- pentose sugar (ribose, deoxyribose)
  - phosphate
  - purine or pyrimidine
    - G,C,A,T
  - nucleoside
    - mono-, di-, tri-
  - phosphodiester bond
  - oligonucleotide
  - polynucleotide
- Watson-Crick model
  - major and minor grooves
  - antiparallel
  - hydrogen bonds
  - complementarity
    - hydrophobic, hydrophilic
- A-DNA, B-DNA, Z-DNA
- RNA
  - rRNA, tRNA, mRNA
- Svedberg coefficient (S)

Processes/Methods

- Replication (F9.1)
- Storage of information(F9.1)
- Expression(F9.1)
  - transcription
  - translation
- Variation (mutation)(F9.1)
- Transformation
  - *Diplococcus pneumoniae*
    - virulent (S)
    - avirulent (R)
    - serotypes (II, III)
    - heat-killed IIIS
      - protease
      - ribonuclease
      - deoxyribonuclease
  - transfection
  - recombinant DNA research
    - insulin
    - human ß-globin gene
    - transgenic mice
      - human growth hormone
  - RNA as genetic material
    - TMV (tobacco mosaic virus)
      - RNA core
      - coat protein
    - Qß phage
      - Qß RNA replicase
    - retroviruses
      - reverse transcription
- X-Ray diffraction
- Hydrogen bonding
- Semiconservative replication

Molecular Hybridization

in situ

reassociation kinetics

half reassociation time (C_{ot})

repetitive sequences

unique (single copy) sequences

Concepts

Central Dogma of Molecular Biology

storage of genetic information

expression of information

variation by mutation

Evidence (direct and indirect) - DNA is genetic material

transformation

transgenic animals

Differential labeling of macromolecules

Circumstantial evidence (DNA in eukaryotes)

DNA distributions

mutagenesis

action spectrum

absorption spectrum

260 nm, 280 nm

Molecular hybridization and sequence complexity

RNA as genetic material (some viruses)

F9.1. Illustration of relationships between DNA, its functions, and related products.

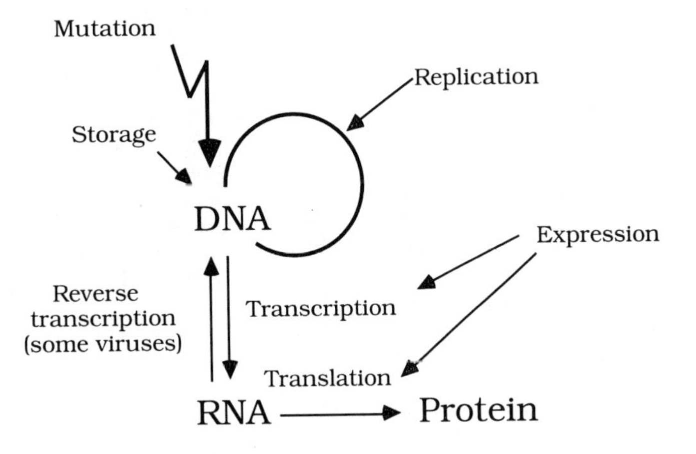

Solutions to Problems and Discussion Questions

1. *Replication* is that process which leads to the production of identical copies of the existing genetic information. Since daughter cells contain essentially exact copies (with some exceptions) of genetic information of the parent cell, and through the production and union of gametes, offspring contain copies (with variation) of parental genetic information, the genetic material must make copies (replicate) of itself. Replication is accomplished during the S phase of interphase.

The genetic material is capable of *expression* through the production of a phenotype. Through transcription and translation, proteins are produced which contribute to the phenotype of the organism. The genetic material must be stable enough to maintain information in "*storage*" from one cell to the next and one organism to the next. Because the genetic material is not "used up" in the processes of transcription and translation, genetic information can be stored and used constantly.

Above, it was stated that the genetic material must be stable enough to store genetic information; however, variation through *mutation* provides the raw material for evolution. The genetic material is capable of a variety of changes, both at the chromosomal and nucleotide levels. See F9.1.

2. Prior to 1940, most of the interest in genetics centered on the transmission of similarity and variation from parents to offspring (transmission genetics). While some experiments examined the possible nature of the hereditary material, abundant knowledge of the structural and enzymatic properties of proteins generated a bias which worked to favor proteins as the hereditary substance. In addition, proteins were composed of as many as twenty different subunits (amino acids) thereby providing ample structural and functional variation for the multiple tasks which must be accomplished by the genetic material. The tetranucleotide hypothesis (structure) provided insufficient variability to account for the diverse roles of the genetic material.

3. Griffith performed experiments with different strains of *Diplococcus pneumoniae* in which a heat-killed pathogen, when injected into a mouse with a live non-pathogenic strain, eventually led to the mouse's death. A summary of this experiment is provided in *Essentials*. Examination of the dead mice revealed living pathogenic bacteria. Griffith suggested that the heat-killed virulent (pathogenic) bacteria transformed the avirulent (non-pathogenic) strain into a virulent strain.

Avery and co-workers systematically searched for the transforming principle originating from the heat-killed pathogenic strain and determined it to be DNA. Taylor showed that transformed bacteria are capable of serving as donors of transforming DNA, indicating that the process of transformation involves a stable alteration in the genetic material (DNA).

4. Nucleic acids contain large amounts of phosphorus and no sulfur whereas proteins contain sulfur and no phosphorus. Therefore the radioisotopes ^{32}P and ^{35}S will selectively label nucleic acids and proteins, respectively.

The Hershey and Chase experiment is based on the premise that the substance injected into the bacterium is the substance responsible for producing the progeny phage and therefore must be the hereditary material. The experiment demonstrated that most of the ^{32}P-labeled material (DNA) was injected while the phage ghosts (protein coats) remained outside the bacterium. Therefore the nucleic acid must be the genetic material.

5. The early evidence would be considered circumstantial in that at no time was there an experiment, like transformation in bacteria, in which genetic information in one organism was transferred to another using DNA. Rather, by comparing DNA content in various cell types (sperm and somatic cells) and observing that the *action* and *absorption* spectra of ultraviolet light were correlated, DNA was considered to be the genetic material. This suggestion was supported by the fact that DNA was shown to be the genetic material in bacteria and some phage. Direct evidence for DNA being the genetic material comes from a variety of observations including gene transfer, which has been facilitated by recombinant DNA techniques.

6. Some viruses contain a genetic material composed of RNA. The tobacco mosaic virus is composed of an RNA core and a protein coat. "Crosses" can be made in which the protein coat and RNA of TMV are interchanged with another strain (Holmes ribgrass). The source of the RNA determines the type of lesion, thus, RNA is the genetic material in these viruses. Retroviruses contain RNA as the genetic material and use an enzyme known as *reverse transcriptase* to produce DNA, which can be integrated into the host chromosome. See F9.1.

7. The structure of deoxyadenylic acid is given below. Linkages among the three components require the removal of water (H_2O).

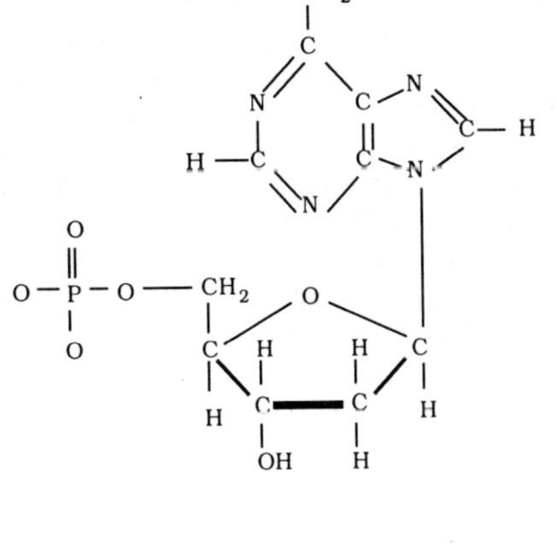

8. Examine the structures of the bases in the *Essentials* text . The other bases would be named as follows:

Guanine: 2-amino-6-oxypurine

Cytosine: 2-oxy-4-aminopyrimidine

Thymine: 2,4-dioxy-5-methylpyrimidine

Uracil: 2,4-dioxypyrimidine

9. The following are characteristics of the Watson-Crick double-helix model for DNA:

The base composition is such that A=T, G=C and (A+G) = (C+T). Bases are stacked, 0.34 nm apart, in a plectonic, antiparallel manner. There is one complete turn for each 3.4 nm, which constitutes 10 bases per turn. Hydrogen bonds hold the two polynucleotide chains together, each being formed by phosphodiester linkages between the sugars and the phosphates. There are two hydrogen bonds forming the A to T pair and three forming the G to C pair. The double helix exists as a twisted structure, approximately 20 nm in diameter, with a topography of major and minor grooves. The hydrophobic bases are located in the center of the molecule while the hydrophilic phosphodiester backbone is on the outside.

10. In addition to creative "genius" and perseverance, model building skills, and the conviction that the structure would turn out to be "simple" and have a natural beauty in its simplicity, Watson and Crick employed the X-ray diffraction information of Franklin and Wilkins, and the base ratio information of Chargaff.

11. Because in double-stranded DNA, A=T and G=C (within limits of experimental error), the data presented would have indicated a lack of pairing of these bases in favor of a single-stranded structure or some other nonhydrogen-bonded structure. Alternatively, from the data it would appear that A=G and T=C which would require purines to pair with purines and pyrimidines to pair with pyrimidines. In that case, the DNA would have contradicted the data from Franklin and Watkins which called for a constant diameter for the double-stranded structure.

12. Three main differences between RNA and DNA are the following:

(1) uracil in RNA replaces thymine in DNA,

(2) ribose in RNA replaces deoxyribose in DNA, and

(3) RNA often occurs as both single- and double-stranded forms, whereas DNA most often occurs in a double-stranded form.

13. The reassociation of separate complementary strands of a nucleic acid, either DNA or RNA, is based on hydrogen bonds forming between A-T (or U) and G-C.

14. In order for hydrogen bonds to form between complementary base pairs, complementary strands of nucleic acids must be in proximity. For a given concentration of nucleic acids, the more copies of a given type of nucleic acid that are present, the higher the likelihood that complementary strands will be close to each other. Conversely, unique sequences of nucleic acids have a lower probability of interacting because there are fewer of them; thus, the likelihood of forming hydrogen bonds (hybridizing) is less.

15.

(1) As shown, the extra phosphate is not normally expected.

(2) In the adenine ring, a nitrogen is at position 8 rather than position 9.

(3) The bond from the C'-1 to the sugar should form with the N at position 9 (N-9) of the adenine.

(4) The dinucleotide is a "deoxy" form, therefore each C-2' should not have a hydroxyl group. Notice the hydroxyl group at C'-2 on the sugar of the adenylic acid.

(5) At the C-5 position on the thymine residue, there should be a methyl group.

(6) At the C'-5 position on the thymidylic acid, there is an extra OH group.

16. Without knowing the exact bonding characteristics of hypoxanthine or xanthine, it may be difficult to predict the likelihood of each pairing type. It is likely that both are of the same class (purine or pyrimidine) because the names of the molecules indicate a similarity. In addition, the diameter of the structure is constant which, under the model to follow, would be expected. In fact, hypoxanthine and xanthine are both purines.

Because there are equal amounts of A, T and H, one could suggest that they are hydrogen bonded to each other; the same may be said for C, G, and X. Given the molar equivalence of erythrose and phosphate, an alternating sugar-phosphate-sugar backbone as in "earth-type" DNA would be acceptable. A model of a triple helix would be acceptable, since the diameter is constant. Given the chemical similarities to "earth-type" DNA it is probable that the unique creature's DNA follows the same structural plan.

17.

(i) The X-ray diffraction studies would indicate a helical structure, for it is on the basis of such data that a helical pattern is suggested. The fact that it is irregular may indicate different diameters (base pairings), additional strands in the helix, kinking or bending.

(ii) The hyperchromic shift would indicate considerable hydrogen bonding, possibly caused by base pairing.

(iii) Such data may suggest irregular base pairing in which purines bind purines (all the bases presented are purines), thus giving the atypical dimensions.

(iv) Because of the presence of ribose, the molecule may show more flexibility, kinking, and/or folding.

While there are several situations possible for this model, the phosphates are still likely to be far apart (on the outside) because of their strong like charges. Hydrogen bonding probably exists on the inside of the molecule and there is probably considerable flexibility, kinking, and/or bending.

18. Since cytosine pairs with guanine and uracil pairs with adenine, the result would be a base substitution of G:C to A:T.

19. Under this condition, the hydrolyzed 5-methyl cytosine becomes thymine.

10

DNA - Replication and Synthesis

Vocabulary: Organization and Listing of Terms

Structures and Substances

DNA polymerase I, II, III

- holoenzyme
  - dimer
- subunits
- *polA1*

5'-nucleotides (F10.1)

3'-nucleotides (F10.1)

Phage øX174

DNA ligase (polynucleotide joining enzyme)

- replisome

Helicase

Single-stranded DNA binding proteins

DNA gyrase (a topoisomerase)

RNA primer

- primase
- free hydroxyl group

DNA ligase

- ligase deficient mutant

ori C

- 9mer, 13mer

*dna*A, *dna*B, *dna*C

Eukaryotic DNA

- four kinds
- semidiscontinuous
  - multirepliconic
- telomeres
  - telomerase
  - ribonucleoprotein (catalytic)

Recombination

- homologous

Heteroduplex DNA molecules

- Holliday structure
- chi form
- *recA*

Processes/Methods

- Replication of DNA (models)
  - semiconservative
    - Meselson and Stahl - 1958
      - *E. coli*
      - equilibrium centrifugation
      - ^{15}N, ^{14}N (in ammonium chloride)
    - Taylor, Woods, and Hughes - 1957
      - *Vicia faba*
        - colchicine
      - ^{3}H-thymidine
      - autoradiography
        - chromatid exchange
  - conservative replication
  - dispersive replication
  - bidirectional (vs. unidirectional)
    - origin of replication, *ori*
    - replicon
      - multiple origins (not random)
    - replication fork
    - continuous, discontinuous
      - leading strand
      - lagging strand
        - Okazaki fragments
- Synthesis of DNA *in vitro*
  - reaction mixture
  - biologically active DNA
  - faithful copying
- Exonuclease activity
- Genetic recombination
  - single-stranded nick
  - endonuclease
- Gene conversion
  - *Neurospora*
  - nonreciprocal

Concepts

- Replication
  - semiconservative
  - complementarity
    - A:T, G:C
    - antiparallel
    - continuous, discontinuous
  - conservative
  - dispersive
- 5' - 3' polarity restrictions
- Supercoiling
- Conditional mutants (F10.2)

F10.1 Shorthand structures for 3' and 5' nucleotides.

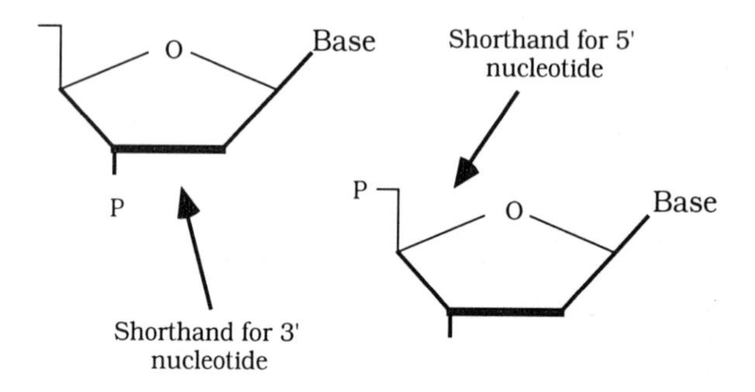

F10.2. Illustration of the influence of a conditional mutant on protein structure and therefore function.

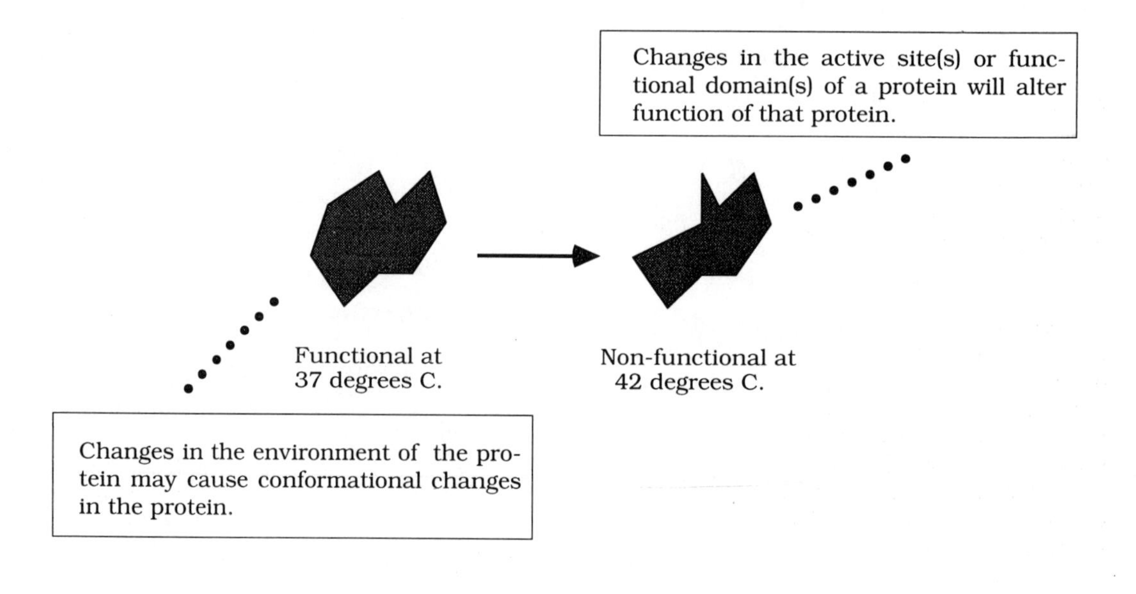

Solutions to Problems and Discussion Questions

1. Refer to Question #1 of the *Insights and Solutions* section of Chapter 10 in *Essentials.* The differences among the three models of DNA replication relate to the manner in which the new strands of DNA are oriented as daughter DNA molecules are produced.

Conservative: In the conservative scheme, the original daughter strand remains as a complete unit and the new DNA double helix is produced as a single unit. The old DNA is completely *conserved.*

Semiconservative: Each daughter strand is composed of one old DNA strand and one new DNA strand. Separation of hydrogen bonds is required.

Dispersive: In the dispersive scheme, the original DNA strand is broken into pieces and the new DNA in the daughter strand is interspersed among the old pieces. Separation of covalent (phosphodiester) bonds is required for this mode of replication.

2. Under a conservative scheme, the first round of replication in ^{14}N medium produces one dense double helix and one "light" double helix in contrast to the intermediate density of the DNA in the semiconservative mode. Therefore, after one round or replication in the ^{14}N medium the conservative scheme can be ruled out.

After one round of replication in ^{14}N under a dispersive model, the DNA is of intermediate density, just as it is in the semiconservative model. However, in the next round of replication in ^{14}N medium, the density of the DNA is between the intermediate and "light" densities. Refer to Question #1 of the *Insights and Solutions* section of Chapter 10 in *Essentials* if you have trouble answering this question.

3. Refer to the *Essentials* text for an illustration of the labeling of *Vicia* chromosomes under a Taylor, Woods, and Hughes experimental design. Notice that only those cells which pass through the S phase in the presence of the ^{3}H-thymidine are labeled and that each double helix (per chromatid) is "half-labeled."

Conservative Replication

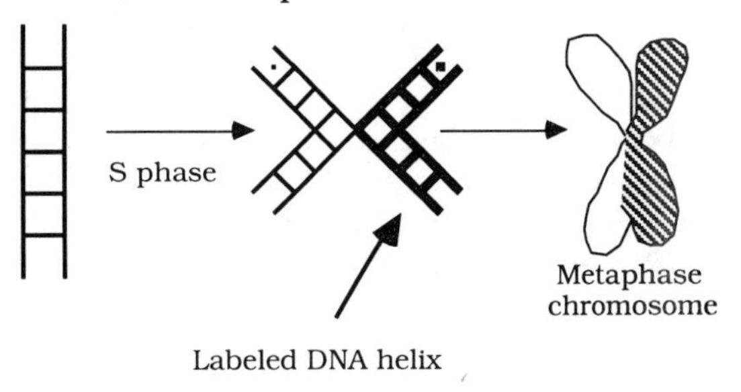

Dispersive Replication

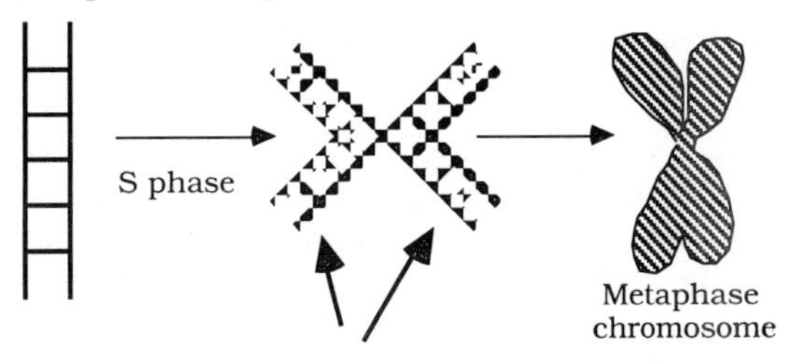

(a) Under a conservative scheme, all of the newly labeled DNA will go to one sister chromatid, while the other sister chromatid will remain unlabeled. In contrast to a semiconservative scheme, the first replicative round would produce one sister chromatid which has label on both strands of the double helix.

(b) Under a dispersive scheme all of the newly labeled DNA will be interspersed with unlabeled DNA. Because these preparations (metaphase chromosomes) are highly coiled and condensed structures derived from the "spread out" form at interphase (which includes the S phase) it is impossible to detect the areas where label is not found. Rather, both sister chromatids would appear as evenly labeled structures.

4. The *in vitro* replication requires a DNA template, a divalent cation (Mg^{++}), and all four of the deoxyribonucleoside triphosphates: dATP, dCTP, dTTP, and dGTP. The lower case "d" refers to the deoxyribose sugar.

5. As stated in the text, *biologically active* DNA implies that the DNA is capable of supporting typical metabolic activities of the cell or organism and is capable of faithful reproduction.

6. ØX174 is a well-studied, single-stranded virus (phage) which can be easily isolated. It has a relatively small DNA genome (5500 nucleotides) which, if mutated, usually alters its reproductive cycle.

7. The *polAI* mutation was instrumental in demonstrating that DNA polymerase I activity was not necessary for the *in vivo* replication of the *E. coli* chromosome. Such an observation opened the door for the discovery of other enzymes involved in DNA replication.

8. All three enzymes share several common properties. First, none can *initiate* DNA synthesis on a template but all can *elongate* an existing DNA strand, assuming there is a template strand as shown in the figure below. Polymerization of nucleotides occurs in the 5' to 3' direction where each 5' phosphate is added to the 3' end of the growing polynucleotide. All three enzymes are large complex proteins with a molecular weight in excess of 100,000 daltons and each has 3' to 5' exonuclease activity. Refer to *Essentials* for a listing of enzyme properties.

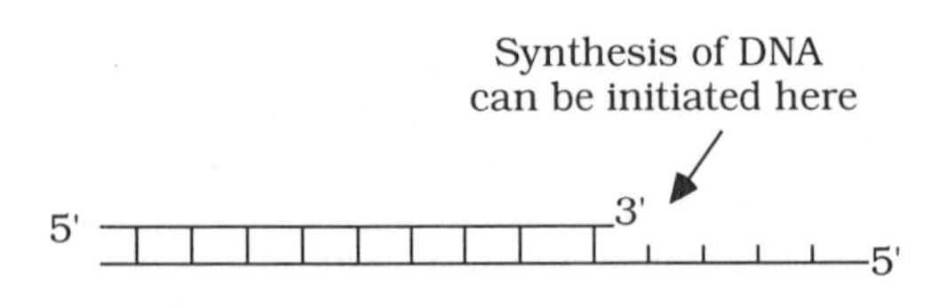

DNA polymerase I:
- 5' to 3' polymerization
- 3' to 5' exonuclease activity
- 5' to 3' exonuclease activity
- present in large amounts
- relatively stable

DNA polymerase II:
- 5' to 3' polymerization
- 3' to 5' exonuclease activity

DNA polymerase III:
- 5' to 3' polymerization
- 3' to 5' exonuclease activity
- essential for replication
- complex molecule
  - seven polypeptide chains
  - possibly 750,000 daltons

9. *Helicase, dnaA* and *single-stranded DNA binding* proteins initially unwind, open, and stabilize DNA at the initiation point. *DNA gyrase*, a DNA topoisomerase, relieves supercoiling generated by helix unwinding. This process involves breaking both strands of the DNA helix.

10. *Okazaki fragments* are relatively short (1000 to 2000 bases in prokaryotes) DNA fragments which are synthesized in a discontinuous fashion on the lagging strand during DNA replication. Such fragments appear to be necessary because template DNA is not available for 5' > 3' synthesis until some degree of continuous DNA synthesis occurs on the leading strand in the direction of the replication fork. The isolation of such fragments provides support for the scheme of replication shown in the *Essentials* text.

DNA ligase is required to form phosphodiester linkages in gaps which are generated when DNA polymerase I removes RNA primer and meets newly synthesized DNA ahead of it.

Notice in the *Essentials* text, the discontinuous DNA strands are ligated together into a single continuous strand.

Primer RNA is formed by RNA primase to serve as an initiation point for the production of DNA strands on a DNA template. None of the DNA polymerases are capable of initiating synthesis without a free 3' hydroxyl group. The primer RNA provides that group and thus can be used by DNA polymerase III.

11. The synthesis of DNA is thought to follow the pattern described in the *Essentials* text. The model involves opening and stabilizing the DNA helix, priming DNA with synthesis with RNA primer, movement of replication forks in both directions which includes elongation of RNA primers in continuous and discontinuous 5' > 3' modes. Okazaki fragments generated in the replicative process are joined together with DNA ligase. DNA gyrase relieves supercoils generated by DNA unwinding.

12. Eukaryotic DNA is replicated in a manner which is very similar to that of *E. coli*. Synthesis is bidirectional, continuous on one strand and discontinuous on the other, and the requirements of synthesis (four deoxyribonucleoside triphosphates, divalent cation, template, and primer) are the same. Okazaki fragments of eukaryotes are about one-tenth the size of those in bacteria.

Because there is a much greater amount of DNA to be replicated and DNA replication is slower, there are multiple initiation sites for replication in eukaryotes in contrast to the single replication origin in prokaryotes. Replication occurs at different sites during different intervals of the S phase. The proposed functions of four DNA polymerases are described in the text.

13. In eukaryotes, the "ends" of chromosomes present a problem in that the 3' end of the lagging strand is an inadequate template, thus, a gap is possible without a unique enzyme called *telomerase*. In some organisms, special telomeric sequences of DNA allow the telomerase to complete replication. The *Tetrahymena* telomerase is known to contain a short piece of RNA which is complementary to the sequence whose synthesis it directs.

14. *Gene conversion* is likely to be a consequence of genetic recombination in which nonreciprocal recombination yields products in which it appears that one allele is "converted" to another. Gene conversion is now considered a result of heteroduplex formation which is accompanied by mismatched bases. When these mismatches are corrected, the "conversion" occurs.

15. (a) Because DNA polymerase III is essential for DNA chain elongation, it is necessary for replication of the *E. coli* chromosome. Thus strains which are mutant for this enzyme must contain conditional mutations. **(b)** The 3' - 5' exonuclease activity is involved in proofreading. Thus proofreading would be hampered in such mutant strains and a higher than expected mutation rate would occur.

11

DNA - Organization in Chromosomes and Genes

Vocabulary: Organization and Listing of Terms

Structures and Substances

Viral chromosomes
- DNA, RNA
  - double-stranded
  - single-stranded
  - often circular
- ϕX174
- lambda (λ)
- T-even bacteriophages
  - packaging

Bacterial chromosomes
- DNA
- double-stranded
- nucleoid
- *E. coli*
  - circular
- DNA-binding proteins
  - HU, H

Mitochondrial DNA (mtDNA)
- semiconservative replication
- plant
- animal
- coding (mtDNA)
  - rRNAs
  - tRNAs
  - respiratory components
- coding (nuclear)
  - imported products
  - antibiotic sensitivity

Chloroplast DNA (cpDNA)
- circular
- double-stranded
- different than nuclear DNA

- coding (cpDNA)
    - rRNAs
    - tRNAs
    - ribulose-1-5-biphosphate carboxylase
        - antibiotic sensitivity
- Eukaryotic chromosomes
    - chromatin
    - mitotic chromosomes
        - condensed chromatin
        - nuclease digesting
            - folded fiber model
                - histones
                - H2A, H2B, H3, H4
                - H1
                - tetramers
                - 200 base pair length
- Heterochromatin/euchromatin
    - unineme model
    - telomeres
        - telomere-associated sequences
        - telomerase
    - centromeres
        - spindle fibers
        - CENs
            - Regions I, II, III
        - Barr body
- Repetitive sequences
    - highly repetitive
        - CEN DNA
        - satellite DNA
        - alphoid family
        - telomeres
            - telomeric DNA sequence
            - telomeric-associated sequence
            - telomerase
    - middle or moderately repetitive
        - SINS
            - *Alu* family
        - LINES
        - VNTRs
    - single copy sequences
        - hnRNA
            - introns
            - exons
- Multigene families
    - globins
        - α family
        - β family
    - pseudogenes
    - ribosomal RNA
        - tandem repeats

nontranscribed spacer regions

nucleolar organizer regions (NORs)

Heterogeneous RNA (hnRNA)

introns (F11.1)

exons (F11.1)

protein domains

LDH receptor proteins

EGF

Promoter regions

TATA box

CCAAT box

enhancers

Processes/Methods

Maternal mode of inheritance

Semiconservative replication

Importing of nuclear-coded gene products

Heterochromatin

few genes

late replicating

position effect

Chromosome banding

C-banding

Q-banding

G-banding

R-banding

Exon shuffling (F11.1)

Concepts

Evolution of cellular organelles

endosymbiont theory

molecular diversity

genetic code

sensitivity to antibiotics

Nuclease digestion

Position effect

Maternal mode of inheritance

Unineme model of chromosomes

Heterochromatin/euchromatin

Colinearity

C value paradox

non-coding DNA

introns

promoter regions

flanking regions

DNA fingerprinting

Exon shuffling and protein domains (F11.1)

Evolution of multigene families

F11.1. Illustration of the relationship between exons and protein domains. *Exon shuffling* is illustrated where exons from gene *A* are found in genes *B* and *C* which each have only one exon (#2 and #3, respectively) in common with gene *A*.

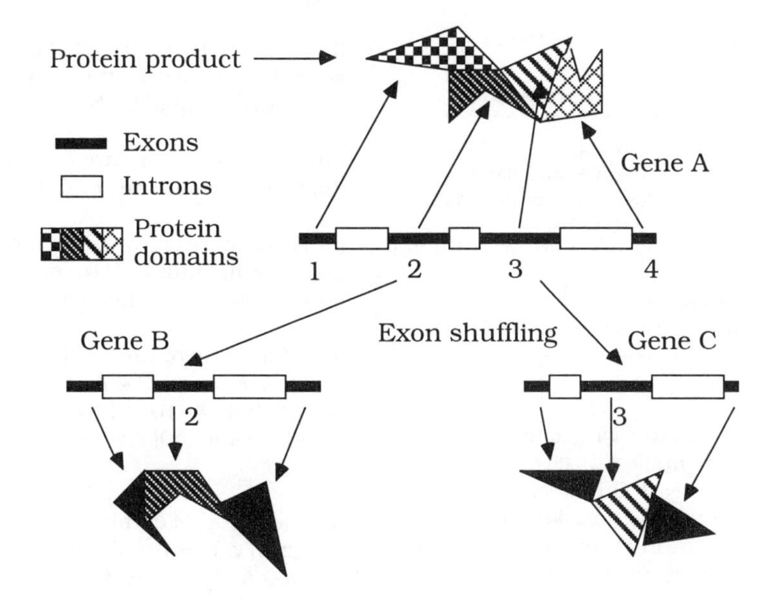

Solutions to Problems and Discussion Questions

1. General similarities and differences:

Viral	*Bacterial*
DNA or RNA	DNA
single-stranded	double-stranded
double-stranded	circular (*E. coli*)
linear	DNA-binding proteins
ring shaped (circular)	genome size >100μm
naked nucleic acid	
genome size < 100μm	

2. General similarities and differences:

mtDNA	*cpDNA*
circular	circular
double-stranded	double-stranded
semiconservative repl.	semiconservative repl.
animal (16 to 18 kb)	>100 kb
plant (>100 kb)	genes(rRNAs, tRNAs, etc.)
genes (rRNAs, tRNAs, etc.)	
diverse (introns in some)	
variations in genetic code	

3. Digestion of chromatin with endonucleases, such as micrococcal nuclease, gives DNA fragments of approximately 200 base pairs or multiples of such. X-ray diffraction data indicated a regular spacing of DNA in chromatin. Regularly spaced bead-like structures (nucleosomes) were identified by electron microscopy. Nucleosomes are octomeric structures of two molecules of each histone (H2A, H2B, H3, and H4) except H1. Between the nucleosomes and complexed with linker DNA is histone H1. A 146-base pair sequence of DNA wraps around the nucleosome and as chromosome condensation occurs a 300-Å fiber is formed. It appears to be composed of 5 or 6 nucleosomes coiled together.

4. *Heterochromatin* is chromosomal material which stains deeply and remains condensed when other parts of chromosomes, euchromatin, are otherwise pale and decondensed. Heterochromatic regions replicate late in S phase and are relatively inactive in a genetic sense because there are few genes present, or if they are present, they are repressed. Telomeres and the areas adjacent to centromeres are composed of heterochromatin.

Examples include regions at the same sites on homologous chromosomes which are genetically "inert" such as telomers and centromeric regions. In addition, certain chromosomal regions have the potential to become heterochromatic such as the X chromosome in female mammals (Barr body) and one haploid set of chromosomes in the mealy bug.

5. (a) Since there are 200 base pairs per nucleosome (as defined in this problem) and 10^9 base pairs, there would be 5×10^6 nucleosomes.

(b) Given that there are 5 nucleosomes per solenoid, there would be 1×10^6 solenoids.

(c) Since there are 5×10^6 nucleosomes and nine histones (including H1) per nucleosome, there must be $9(5 \times 10^6)$ histone molecules: 4.5×10^7.

(d) Since there are 10^9 base pairs present and each base pair is 3.4 Å the overall length of the DNA is 3.4×10^9 Å. Dividing this value by the packing ratio (50) gives 6.8×10^7.

6. The first part of this problem is to convert all of the given values to cubic Å remembering that 1 μm = 1000 Å. Using the formula πr^2 for the area of a circle and $4/3\ \pi r^3$ for the volume of a sphere, the following calculations apply:

Volume of DNA: $3.14 \times 10Å \times 10Å \times (50 \times 10^4Å) = 1.57 \times 10^8 Å^3$

Volume of capsid: $4/3\ (3.14 \times 400Å \times 400Å \times 400Å) = 2.67 \times 10^8 Å^3$

Because the capsid head has a greater volume than the volume of DNA, the DNA will fit into the capsid.

7. *Exon shuffling* is a term used to describe the likely phenomenon whereby the exons of genes, each coding for functional domains of proteins, can be "shuffled" or rearranged to facilitate the evolution of a new protein. Rather than each protein evolving "from scratch," it is suggested that they are built from previously-evolved gene segments, exons. Several lines of evidence support the "exon shuffling" model:

(a) exon size is consistent with the size of functional domains of proteins,

(b) exon/intron boundaries often match up with functional domains of proteins,

(c) the major recombinational events occur in the introns, thus leaving the exons as potentially mobile units, and

(d) striking sequence homology exists between exons from genes which code for different proteins. See F11.1.

8. While the β-globin gene family is relatively large (60kb) sequence and restriction analyses show that it is composed of six genes, one is a pseudogene and therefore does not produce a product.

The five functional genes each contain two similarly-sized introns which when included with non-coding flanking regions (5' and 3'), and spacer DNA between genes, accounts for the 95% mentioned in the question.

9. The *rRNA* gene family is well characterized in a variety of organisms. It generally consists of tandem repeats of three molecules in the following order: 18S, 5.8S, and 28S rRNA. There are substantial regions between each gene which are transcribed but subsequently removed. Thus, the initial transcript is processed similar to the removal of introns. Between each three-gene unit in the cluster is a non-transcribed spacer DNA sequence.

The transcription unit varies in size from 7kb to 13kb in humans, the variation coming from the different sized spacer DNAs. The basic family unit, including the non-transcribed spacers, contains about 43kb of tandemly repeated DNA.

In humans the rRNA gene clusters are located on the ends of chromosomes #13, #14, #15, #21, and #22 while in *Drosophila* they are located on the X and Y chromosomes. The chromosomal regions which include the DNA coding for rRNAs are called *nucleolus organizer regions* (NORs).

10. First, a "C value" is the amount of DNA contained in a haploid genome. The *C value paradox* recognizes that with evolutionary divergence there has been a dramatic divergence in the amount of DNA among different taxa. It is likely that with the phylogenetic divergence there is not a commensurate requirement for *that many* different and additional genes. Indeed, some organisms which are quite closely related phylogenetically have vastly different DNA contents. So what is this "extra" DNA doing in various genomes?

Until recently, answers to this question were vague and speculative. However, with recent advances in molecular biology, especially cloning and sequencing of DNA, it has been determined that much of this "extra" DNA is found in heterochromatic regions, introns, regions flanking genes, and intergenic spacer regions. While the nature of large amounts of "non-coding" DNA has been described, the question that still remains is "why do different taxa retain such different amounts of this "extra" DNA?" The C value paradox is being answered at a "within taxon" level but not too well at the "among taxa" level.

12

Mutation, Repair, and Transposable Elements

Vocabulary: Organization and Listing of Terms

Structures and Substances

Somatic cells

Gamete forming cells

- germ line

5-Bromouracil

2-Amino purine

Acridine orange

Proflavin

Mustard gas

Ethylmethane sulfonate

ABO antigens

- H substance
  - glycosyltransferase

Dystrophin

Pyrimidine dimers

uvr gene product

DNA polymerase I

DNA ligase

mutH, L, S and *U*

recA

Photoreactivation enzyme

Heterokaryon

Transposable elements

- insertion sequences
- Ds (Dissociation)
- Ac (Activator)
- transposon
- copia
- terminal repeats (DTR, ITR)
  - P, I elements
- *Alu* family
- SINE
- LINE

Processes/Methods

- Variation by mutation
  - spontaneous
    - background radiation
      - cosmic sources
      - mineral sources
      - ultraviolet light
    - rates
  - induced
- Molecular basis
  - base substitution or point mutations
    - transition
    - transversion
    - frameshift
  - tautomeric shifts (forms)
    - keto, enol
  - base analogues
    - 5-bromouracil
    - 2-amino purine
  - reverse mutation
  - alkylation
    - mustard gases
    - ethylmethane sulfonate
    - 6 - ethyl guanine
  - frameshift mutations
    - acridine dyes (acridines)
      - acridine orange
    - proflavin
- Detection
  - bacteria and fungi
    - minimal medium
    - complete medium
    - prototrophs
    - auxotrophs
  - *Drosophila*
    - attached-X procedure
  - plants
    - biochemical
    - visual observation
    - tissue culture
  - humans
    - pedigree analysis
    - cell culture (*in vitro*)
    - ABO antigens
      - H substance modification
    - muscular dystrophy
      - myopathy
      - Duchenne muscular dystrophy
      - Becker muscular dystrophy
        - dystrophin

FMR-1

fragile-X

myotonic dystrophy

Huntington Disease

trinucleotide DNA repeats

spinobulbar muscular atrophy (Kennedy disease)

Ames test

Salmonella typhimurium

Repair

high-energy radiation

X-rays

gamma radiation

cosmic radiation

free radicals

reactive ions

intensity of dose

roentgen

doubling dose of radiation

ultraviolet radiation (260nm)

pyrimidine dimers

T-T, C-C, T-C

photoreactivation

photoreactivation enzyme (PRE)

excision repair

uvr gene product

DNA polymerase I

DNA ligase

proofreading and mismatch repair

strand discrimination

DNA methylation

mutH, L, S and *U*

recombinational repair

recA

rescue operation

SOS response

xeroderma pigmentosum (XP)

unscheduled DNA synthesis

photoreactivation enzyme

heterokaryon

somatic cell genetics

complementation

Site-directed mutagenesis

Genetic transposition

Hybrid dysgenesis

Concepts

Mutation - basis of organismic diversity (F12.1)

chromosomal aberrations

gene mutations

somatic (F12.2)

germ line (F12.2)

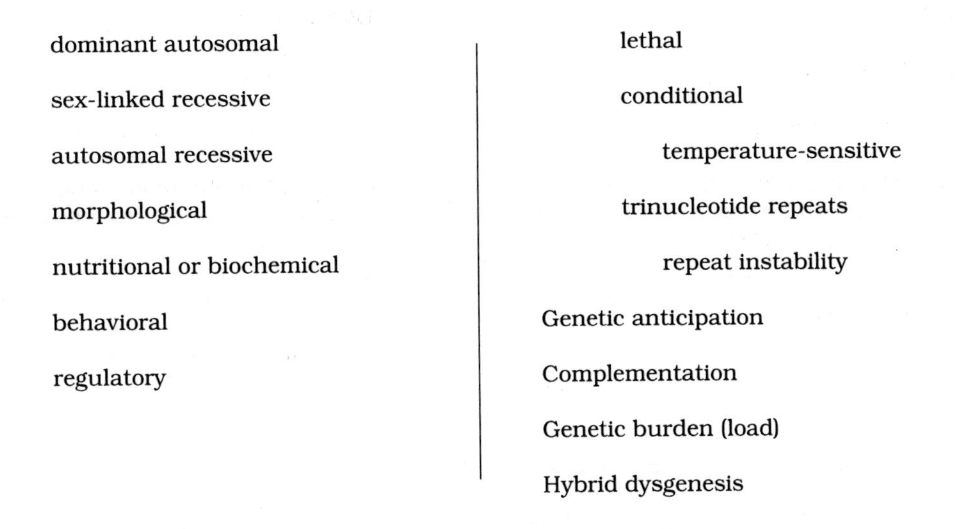

F12.1. Graphic representation of the relationship of mutation to Darwinian evolutionary theory. Mutation provides the original source of variation on which natural selection operates.

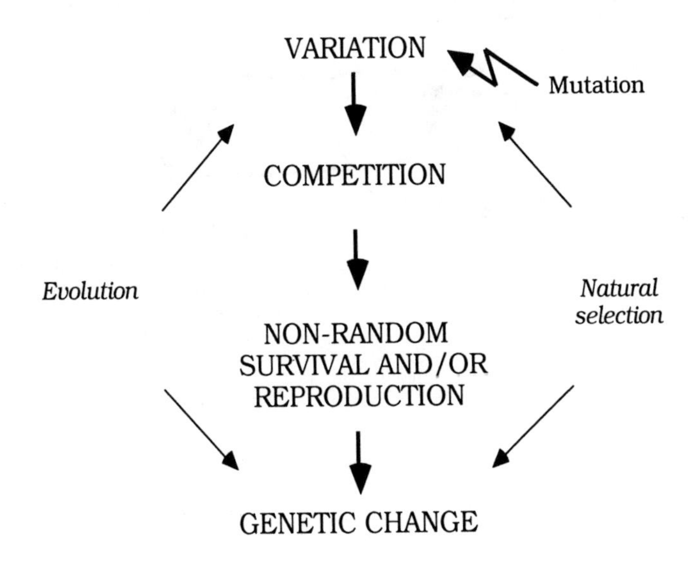

F12.2. Illustration of the difference between somatic and germ-line mutation. Somatic mutations are not passed to the next generation whereas those in the germ line may be passed to offspring.

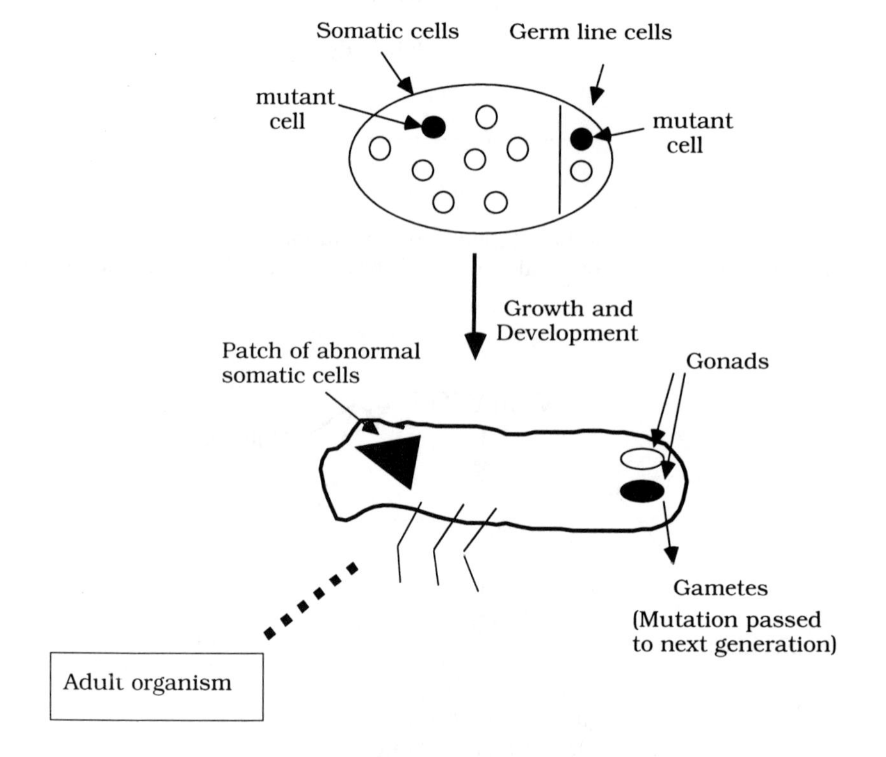

Solutions to Problems and Discussion Questions

1. The term *chromosomal mutation* refers to changes in chromosome number or structure, such as duplications, deletions, inversions, and translocations. A *gene mutation* is a change in the nucleotide sequence of a single gene.

2. A functional sequence of nucleotides, a gene, is likely to be the product of perhaps a billion or so years of evolution. Each gene and its product function in an environment which has also evolved, or co-evolved.

A coordinated output of each gene product is required for life. Deviations from the norm, caused by mutation, are likely to be disruptive because of the complex and interactive environment in which each gene product must function. However, on occasion a beneficial variation occurs.

3. A diploid organism possesses at least two copies of each gene (except for "hemizygous" genes) and in most cases, the amount of product from one gene of each pair is sufficient for production of a normal phenotype. Recall that the condition of "recessive" is defined by the phenotype of the heterozygote. If one unit of output from the normal gene gives the same phenotype as in the normal homozygote, where there are two units of output, the allele is considered "recessive."

	Phenotype, if mutant is:	
Genotypes	**recessive**	**dominant**
wild/wild	wild	wild
wild/mutant	wild	mutant
mutant/mutant	mutant	mutant

4. Let *II* indicate a mutagenized second chromosome

Females *Cy L/Pm* X males *II/II*

F_1 males *Cy L/II* X *Cy L/Pm*

(individual crosses)

F_2 females *Cy L/II* X *Cy L/II*

(individual crosses)

F_3 genotypes:

Cy L/Cy L	(lethal)
Cy L/II	(Curly, Lobe)
II/II	(dies if recessive lethal)

The *II/II* class will be present in crosses where no recessive lethal was introduced. If a recessive lethal had been introduced, only Curly/Lobe flies would be seen in the cultures. In addition, recessive morphological mutations will be expressed in the *II/II* offspring.

5. Watson and Crick recognized that various tautomeric forms, caused by single proton shifts, could exist for the nitrogenous bases of DNA. Such shifts could result in mutations by allowing hydrogen bonding of normally noncomplementary bases. As stated in the text, important tautomers involve keto-enol pairs for thymine and guanine, and amino-imino pairs for cytosine and adenine.

6. Frameshift mutations are likely to change more than one amino acid in a protein product because as the reading frame is shifted, new codons are generated. In addition, there is the possibility that a nonsense triplet could be introduced, thus causing premature chain termination. If a single pyrimidine or purine has been substituted, then only one amino acid is influenced.

7. *Photoreactivation* can lead to repair of UV-induced damage. An enzyme, photoreactivation enzyme, will absorb a photon of light to cleave thymine dimers. *Excision repair* involves the products of several genes, DNA polymerase I, and DNA ligase to clip out the UV-induced dimer, fill in, and join the phosphodiester backbone in the resulting gap. The excision repair process can be activated by damage which distorts the DNA helix.

Recombinational repair is a system which responds to DNA that has escaped other repair mechanisms at the time of replication. If a gap is created on one of the newly synthesized strands, a "rescue operation or SOS response" allows the gap to be filled. Many different gene products are involved in this repair process. In SOS repair, the normal proofreading of DNA polymerase III is suppressed and this therefore is called an "error-prone system."

8. Because mammography involves the use of X-rays and X-rays are known to be mutagenic, it has been suggested that frequent mammograms may do harm.

9. In *excision repair* a small section of DNA is removed, and subsequently "filled-in" by DNA polymerase activity. Such represents "unscheduled DNA synthesis." One can determine complementation groupings by placing each heterokaryon giving a "0" into one group and those giving a "+" into a separate group. For instance, *XP1* and *XP2* are placed into the same group because they do not complement each other. However, *XP1* and *XP5* do complement ("+") therefore they are in a different group. Completing such pairings allows one to determine the following groupings:

XP1	*XP4*	*XP5*
XP2		*XP6*
XP3		*XP7*

The groupings (complementation groups) indicate that there are at least three "genes" which form products necessary for unscheduled DNA synthesis. All of the cell lines which are in the same complementation group are defective in the same product.

10. Each individual arises from the union of two gametes. If there are 100,000 genes per genome (haploid) then the number of new mutations per individual would be as follows:

$$2(5 \text{ X } 10^{-5})(1 \text{ X } 10^{5}) = 10$$

Assuming 4.3 X 10^9 individuals, there would be 4.3 X 10^{10} new mutations in the current populace.

11. The sequence of bases in a wild type gene provides a code which is eventually translated into the amino acid sequence of a protein. If an insertion sequence or transposable element is inserted into that wild type sequence, the original coding is disrupted, thus leading to a modified (mutant) gene product. In addition, insertion sequences and transposable elements may carry DNA sequences which influence a variety of transcriptional and translational activities.

13

Storage and Expression of Genetic Information

Vocabulary: Organization and listing of terms

Structures and Substances

Codon, anticodon

Messenger RNA

Polynucleotide phosphorylase

- random assembly of nucleotides

Homopolymer codes

- RNA homopolymers
- RNA heteropolymers

N-formylmethionine (fmet)

MS2 phage

Polypeptide

Ribosome

- rRNA
- 5SRNA
- 5.8SRNA
- subunits
- ribosomal proteins
- monosome (70S, 80S)
- small subunit
- large subunit
    - maturation
    - chromosomal origins

Ribosome complex

- peptidyl (P site)
- aminoacyl (A site)
- peptidyl transferase

Adaptor molecules (transfer RNAs - tRNA)

Messenger RNA (mRNA)

RNA polymerase (*E. coli*)

- holoenzyme (α, β, β', σ)
- nucleoside triphosphates (NTPs)
- nucleoside monophosphates (NMPs)
- nucleotidcs

RNA polymerase (eukaryotic) - I, II, III

- promoters (promoter sequences)
- template, partner strands

Consensus sequences

- TATA
- -25 Golberg-Hogness or TATA box
- CCAAT box
- transcription factors
- enhancers

tRNA (4S)

- codon/anticodon
- precursors
    - modified bases
- cloverleaf model
    - paired and unpaired loops
    - ...pCpCpA (3')
    - ...pG (5')
- aminoacyl tRNA synthetases
    - charging
    - activated form
        - (aminoacyladenylic acid)

Initiation factors

- formylmethioninyl - tRNA
- initiation complex
- Shine-Delgarno sequence

Elongation

- peptidyl site
- aminoacyl site
- GTP-dependent release factors
- protein elongation factors

Polyribosomes (polysomes)

Heterogeneous RNA (hnRNA)

- pre-mRNAs
- split genes (intervening sequences)
    - introns
    - exons
    - heteroduplexes
- poly-A
- cap (7mG)
    - 5' to 5'
- ribozyme
- splicosome
- small nuclear ribonucleoprotein (snRNP)
- isoforms

Processes/Methods

Transcription, Translation

Cell-free protein-synthesizing system

Triplet binding assay

Repeating copolymers

Chain elongation (5' to 3')

Chain termination

- termination factor (ρ)

Translation

- tRNA charging
  - aminoacyl tRNA synthetases
  - charging
  - activated form
    - (aminoacyladenylic acid)
- chain initiation
- chain elongation
  - translocation
- chain termination
  - UAG, UAA, UGA

RNA processing

- split genes
- post-transcriptional changes
  - poly-A (3')
  - cap (5')
- pre-mRNA
- hnRNA
  - Groups I-IV
  - self-excision
  - lariat-like structure

Overlapping genes

- multiple initiation

Concepts

Genetic code

- triplet codon
  - unambiguous
  - degenerate
  - ordered
  - commaless
  - nonoverlapping
  - universal
    - exceptions
  - codon assignments
    - artificial mRNAs
      - homopolymers
      - copolymers
- % amino acid incorporation
  - triplet binding assay
  - repeating copolymers
  - confirmation of codon assignments
    - MS2 sequencing
    - colinear relationship
- punctuation
  - start
    - initiator codon
    - N-formylmethionine (fmet)
    - AUG, GUG (rare)

stop

UAA, UAG, UGA

nonsense mutation

amber, ochre, opal

universal, with exceptions

yeast, human mitochondrial DNA

wobble hypothesis

pattern of degeneracy

Information flow (F13.1)

transcription

RNA polymerases (I, II, III)

consensus sequences

translation (F13.2)

RNA processing

RNA splicing

intervening sequences

alternative splicing

isoforms

F13.1 Illustration of the processes, transcription and translation, involved in protein synthesis. Such relationships are often called the **Central Dogma**.

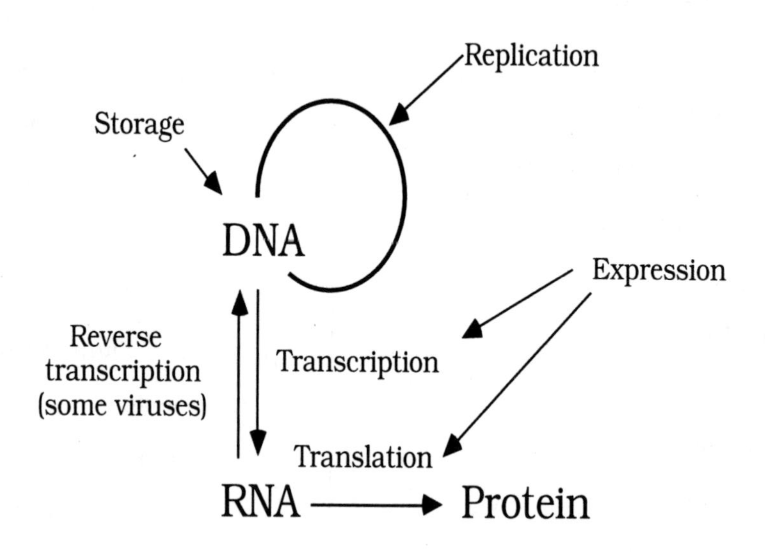

F13.2. Polarity constraints associated with simultaneous transcription and translation. The RNA polymerase is moving downward (**bold arrow**) in this drawing.

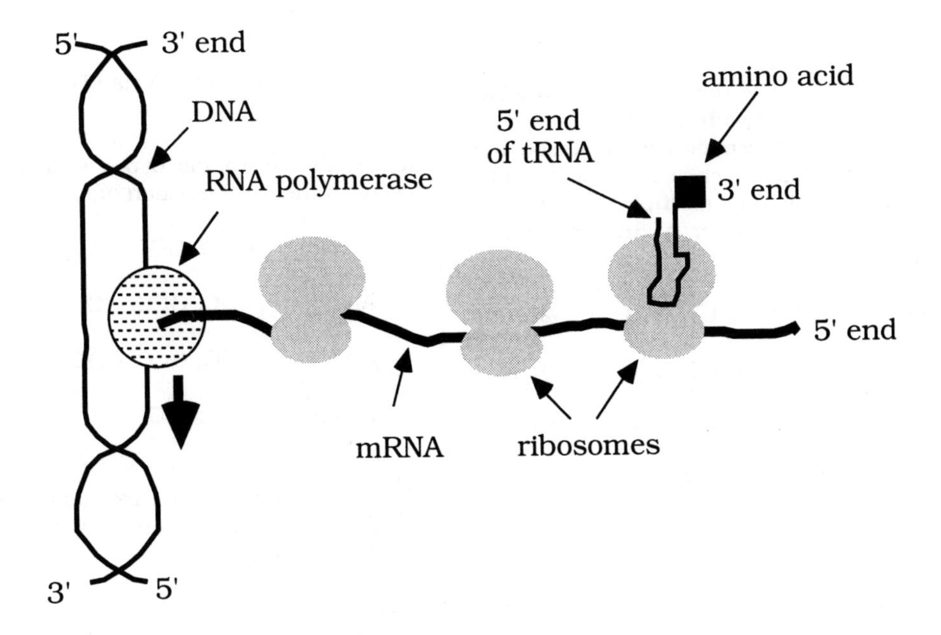

Solutions to Problems and Discussion Questions

1. No. The term "reading frame" refers to the number of bases contained in each codon. The reason that (+++) or (- - -) restored the reading frame *is because* the code is triplet. By having the (+++) or (- - -), the translation system is "out of phase" until the third "+" or "-" is encountered. If the code contained six nucleotides (a sextuplet code), then the translation system is "out of phase" until the sixth "+" or "-" is encountered.

2. (a) The way to determine the fraction which each triplet will occur with a random incorporation system is to determine the likelihood that each base will occur in each position of the codon (first, second, third), then multiply the individual probabilities (fractions) for a final probability (fraction).

GGG = 3/4 X 3/4 X 3/4 = 27/64

GGC = 3/4 X 3/4 X 1/4 = 9/64

GCG = 3/4 X 1/4 X 3/4 = 9/64

CGG = 1/4 X 3/4 X 3/4 = 9/64

CCG = 1/4 X 1/4 X 3/4 = 3/64

CGC = 1/4 X 3/4 X 1/4 = 3/64

GCC = 3/4 X 1/4 X 1/4 = 3/64

CCC = 1/4 X 1/4 X 1/4 = 1/64

(b) Glycine:

GGG and one G_2C (adds up to 36/64)

Alanine:

one G_2C and one C_2G (adds up to 12/64)

Arginine:

one G_2C and one C_2G (adds up to 12/64)

Proline:

one C_2G and CCC (adds up to 4/64)

(c) With the wobble hypothesis, variation can occur in the third position of each codon.

Glycine: GGG, GGC

Alanine: CGG, GCC, CGC, GCG

Arginine: GCG, GCC, CGC, CGG

Proline: CCC,CCG

3. First, compute the frequency (percentages would be easiest to compare) for each of the random codons.

For 4/5 C: 1/5 A:

CCC= 4/5 X 4/5 X 4/5 = 64/125 (51.2%)

C_2A = 3(4/5 X 4/5 X 1/5) = 48/125 (38.4%)

CA_2 = 3(4/5 X 1/5 X 1/5) = 12/125 (9.6%)

AAA= 1/5 X 1/5 X 1/5 = 1/125 (0.8%)

For 4/5 A: 1/5 C:

AAA= 4/5 X 4/5 X 4/5 = 64/125 (51.2%)

A_2C = 3(4/5 X 4/5 X 1/5) = 48/125 (38.4%)

AC_2 = 3(4/5 X 1/5 X 1/5) = 12/125 (9.6%)

CCC= 1/5 X 1/5 X 1/5 = 1/125 (0.8%)

Proline: C_3, and one of the C_2A triplets

Histidine: one of the C_2A triplets

Threonine: one C_2A triplet, and one A_2C triplet

Glutamine: one of the A_2C triplets

Asparagine: one of the A_2C triplets

Lysine: A_3

4. As in the previous problem, the procedure is to find those sequences which are the same for the first two bases but which vary in the third base. Given that AGG = arg, then information from the AG copolymer indicates that AGA also codes for arg and GAG must therefore code for glu.

Coupling this information with that of the AAG copolymer, GAA must also code for glu, and AAG must code for lys.

5. The basis of the technique is that if a trinucleotide contains bases (a codon) which are complementary to the anticodon of a charged tRNA, a relatively large complex is formed which contains the ribosome, the tRNA, and the trinucleotide. This complex is trapped in the filter whereas the components by themselves are not trapped. If the amino acid on a charged, trapped tRNA is radioactive, then the filter becomes radioactive.

6. Apply the most conservative pathway of change.

gly (GGA,G) → agr (AGA,G) → thr (ACA,G); ser (AGU,C); ileu (AUA)

gly (GGA,G) → glu (GAA,G) → val (GUA,G); ala (GCA,G)

7. The central dogma of molecular genetics and to some, all of biology, states that DNA produces, through transcription, RNA, which is "decoded" (during translation) to produce proteins. See F13.1 for a graphic description.

8. A functional polyribosome will contain the following components:

mRNA, charged tRNA, large and small ribosomal subunits, elongation and perhaps initiation factors, peptidyl transferase, GTP, Mg^{++}, nascent proteins, possibly GTP-dependent release factors.

9. Transfer RNAs are "adaptor" molecules in that they provide a way for amino acids to interact with sequences of bases in nucleic acids. Amino acids are specifically and individually attached to the 3' end of tRNAs which possess a three-base sequence (the anticodon) to base-pair with three bases of mRNA. Messenger RNA, on the other hand, contains a copy of the triplet codes which are stored in DNA. The sequences of bases in mRNA interact, three at a time, with the anticodons of tRNAs.

10. It was reasoned that there would not be sufficient affinity between amino acids and nucleic acids to account for protein synthesis. For example, acidic amino acids would not be attracted to nucleic acids. With an adaptor molecule, specific hydrogen bonding could occur between nucleic acids, and specific covalent bonding could occur between an amino acid and a nucleic acid tRNA.

11. The sequence of base triplets in mRNA constitutes the sequence of codons. A three-base portion of the tRNA constitutes the anticodon.

12. Since there are three nucleotides which code for each amino acid, there would be 423 code letters (nucleotides), 426 including a termination codon. This assumes that other features, such as the polyA tail, the 5'cap, and non-coding leader sequences are omitted. Dividing 20 by 0.34, gives the number of nucleotides (about 59) occupied by a ribosome. Dividing 59 by three gives the approximate number of triplet codes: approximately 20.

13. The steps involved in tRNA charging are outlined in *Essentials*-Fig.13.11. An amino acid in the presence of ATP, Mg^{++}, and a specific aminoacyl synthetase produces an amino acid-AMP enzyme complex (+ PP_i). This complex interacts with a specific tRNA to produce the aminoacyl tRNA.

14. One can conclude that the amino acid is not involved in recognition of the anticodon.

15.

(a)

```
      G - C       C - C
     /     \     /     \
A — G        G           G
|   |        |            \
|   |        |             A
|   |        |            /
U — C        C      U    C
     \ A—U  /    \     /
```

(b) TCCGCGGCTGAGATGA (use complementary bases, substituting T for U)
(c) GCU
(d) Assuming that the AGG... is the 5' end of the mRNA, then the sequence would be

arg-arg-arg-leu-tyr

16.

(a) #1: *nonsense mutation*
#2: *frameshift mutation*
#3: *missense mutation*

(b) #1: mutation in third position to A or G
#2: removal of a G in the UGG triplet (trp)
#3: change from U to C

(c) promoter or enhancer, although many posttranscriptional alterations are possible

14

Proteins: The End Product of Genes

Vocabulary: Organization and Listing of Terms

Structures and Substances

- Garrod
  - alkaptonuria
    - homogentisic acid
  - phenylketonuria (PKU)
    - phenylalanine hydroxylase
    - phenylalanine
    - phenylpyruvic acid
- Polypeptide chains
  - quaternary structure
  - hemoglobin
    - heme group
    - globin portion
      - HbA
      - HbA_2
      - HbF
      - Gower 1
      - HbC
    - chains
      - chromosome 16
        - alpha (α)
        - zeta (ζ)
      - chromosome 11
        - beta (β)
        - delta (δ)
        - epsilon (ε)
        - gamma ($^{G}\gamma$),($^{A}\gamma$)
  - sickle-cell hemoglobin
    - β chain
    - 6th position (glutamic acid —> valine)
    - molecular disease
- Polypeptide/Proteins
  - carboxyl group (C-terminus)
  - amino group (N-terminus)

R (radical group)

non-polar (hydrophobic)

polar (hydrophilic)

negatively, positively charged

peptide bond

dipeptide, tripeptide

structure

primary (I°)

secondary (II°)

β-pleated sheet

tertiary (III°)

quaternary (IV°)

Posttranslational modification

end modifications

kinases

glycoproteins

trimming

signal sequence

targeting

metalloproteins

collagen

tropocollagen

procollagen

Ehlers-Danlos syndrome

procollagen peptidase

osteogenesis imperfecta

type I , type II

Marfan's syndrome

Processes/Methods

Enzymes

energy of activation

active site, catalytic site

catabolic, anabolic, amphibolic

Electrophoretic migration

starch gel, two-dimensional

anode, cathode

Fingerprinting technique

Posttranslational modification

Concepts

One-gene: one-enzyme

One-gene: one-protein

sickle-cell anemia, sickle-cell trait

One-gene: one-polypeptide chain

Pathway analysis (F14.1)

Neurospora

Colinear relationships (nucleotides, proteins)

Protein structure (F14.2)

Posttranslational modification

Protein targeting

F14.1. An illustration of the relationship between a metabolic block (caused by a mutant gene) and a metabolic pathway. Notice that one mutant gene causes the pathway to be blocked at a specific place.

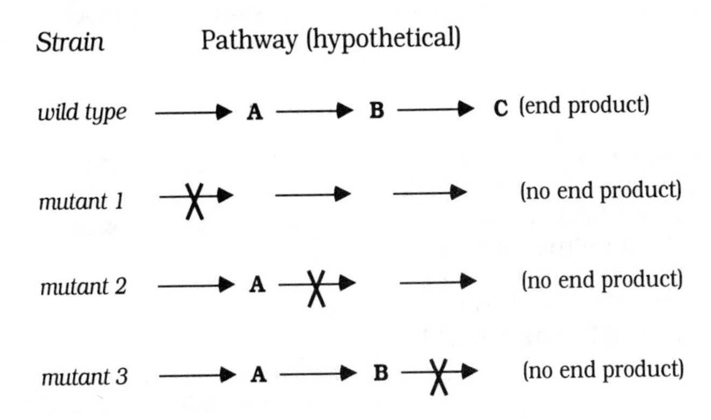

F14.2. Below is a simple sketch of several folding aspects of a protein.

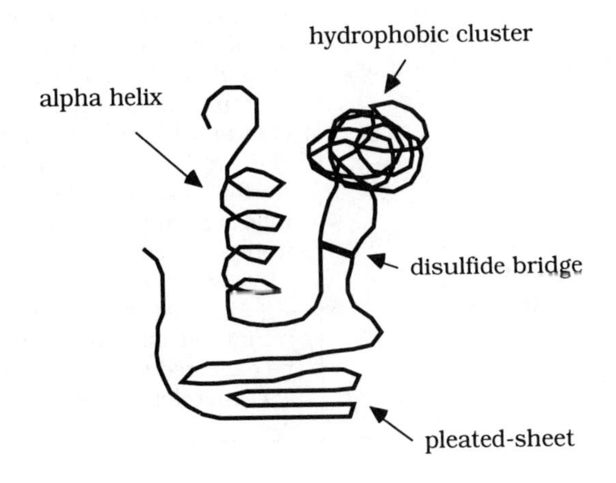

Solutions to Problems and Discussion Questions

1. Phenylalanine is an amino acid which, like other amino acids, is required for protein synthesis. While too much phenylalanine and its derivatives cause PKU, too little will restrict protein synthesis.

2. (a) In this cross, two gene pairs are operating because the F_2 ratio is a modification of a 9:3:3:1 ratio, which is typical of a dihybrid cross. If one assumes that homozygosity for either or both of the two loci gives white, then let strain *A* be, *aaBB* and strain *B*, *AAbb*. The F_1 is *AaBb* and pigmented (purple). The typical F_2 ratio would be as follows:

9/16	*A_B_*	purple
3/16	*aaB_*	white
3/16	*A_bb*	white
1/16	*aabb*	white

If a pathway exists which has the following structure, then the genetic and biochemical data are explained.

aa **bb**

X ---|-----> **Y** ---|-----> **purple**
(white) **(white)** **pigment**

(b) For this condition, with the pink phenotype present, leave the symbols the same however change the Y compound such that when accumulated, a pink phenotype is produced:

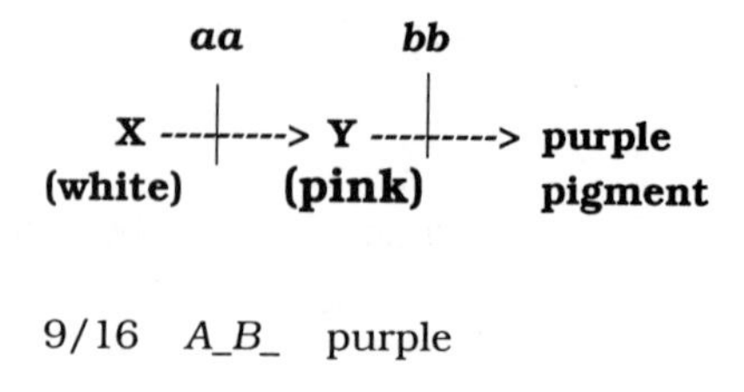

9/16	*A_B_*	purple
3/16	*aaB_*	white
3/16	*A_bb*	**pink**
1/16	*aabb*	white

3. The best way to approach these types of problems, especially when the data are organized in the form given, is to realize that the substance (supplement) which "repairs" a strain, as indicated by a (+), is *after* the metabolic block for that strain. In addition, and most importantly, is that the substance which "repairs" the highest number of strains either *is the end product* or is *closest to* the end product.

Looking at the table, notice that the supplement tryptophan "repairs" all the strains. Therefore it must be at the end of the pathway or at least after all the metabolic blocks (defined by each mutation). Indole "repairs" the next highest number of strains (3), therefore it must be second from the end. Indole glycerol phosphate "repairs" two of the four strains so it is third from the end. Anthranilic acid "repairs" the least number of strains, so it must be early (first) in the pathway.

Minimal medium is void of supplements, and mutant strains involving this pathway would not be expected to grow (or be "repaired"). The pathway therefore would be as follows:

AA----->IGP----->I----->TRY

To assign the various mutations to the pathway, keep in mind that if a supplement "repairs" a given mutant, the supplement must be after the metabolic block. Applying this rationale to the above pathway, the metabolic blocks are created at the following locations.

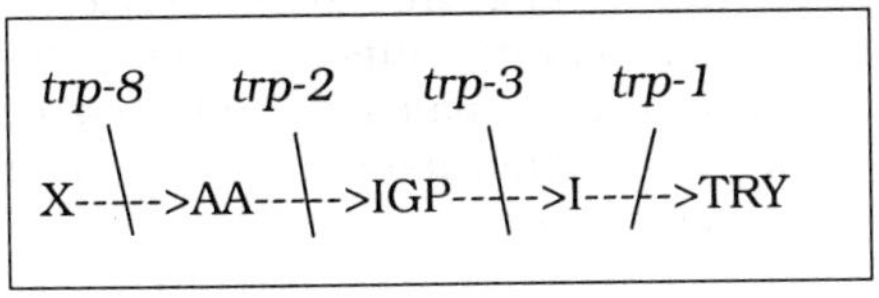

```
trp-8     trp-2     trp-3     trp-1
X---\--->AA---|--->IGP---\--->I---/--->TRY
```

4. In general, the rationale for working with a branched chain pathway is similar to that stated in the previous problem. Since thiamine "repairs" each of the mutant strains, it must, as stated in the problem, be the final synthetic product.

Remembering the "one-gene: one-enzyme" statement, each metabolic block should only occur in one place, so even though pyrimidine and thiazole supplements each "repair" only one strain each, they will not occupy the same step; rather a branched pathway is suggested. Consider that pyrimidine and thiazole are products of distinct pathways and that both are needed to produce the end product, thiamine, as indicated below:

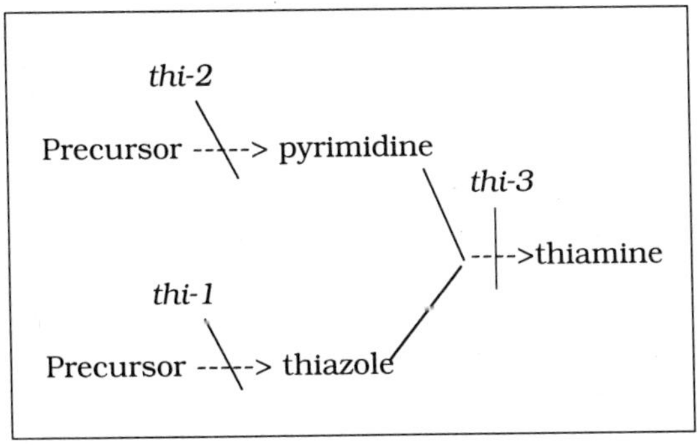

```
          thi-2
Precursor --\--> pyrimidine
                           \   thi-3
                            \---|-->thiamine
          thi-1             /
Precursor --\--> thiazole
```

5. The fact that enzymes are a subclass of the general term *protein*, a *one-gene:one-protein* statement might seem to be more appropriate. However, some proteins are made up of subunits, each different type of subunit (polypeptide chain) being under the control of a different gene. Under this circumstance, the *one-gene:one-polypeptide* might be more reasonable.

It turns out that many functions of cells and organisms are controlled by stretches of DNA that either produce no protein product (operator and promoter regions, for example) or possess more than one function as in the case of overlapping genes and differential mRNA splicing. A simple statement regarding the relationship of a stretch of DNA to its physical product is difficult to formulate.

6. Sickle-cell anemia is coined a *molecular* disease because it is well understood at the molecular level; at the level of a base change in DNA which leads to an amino acid change in the β chain of hemoglobin. It is a *genetic* disease in that it is inherited from one generation to the next. It is not contagious as might be the case of a disease caused by a microorganism. Diseases caused by microorganisms may not necessarily follow family blood lines whereas genetic diseases do.

7. *Colinearity* refers to the sequential arrangement of subunits, amino acids and nitrogenous bases in proteins and DNA, respectively. Sequencing of genes and products in MS2 phage and studies on mutations in the *A* subunit of the *tryptophan synthetase* gene indicate a colinear relationship.

8. As stated in the text, the four levels of protein structure are the following:

Primary: the sequence of amino acids. This sequence determines the higher level structures.

Secondary: α–helix and β-pleated-sheet structures generated by hydrogen bonds between components of the peptide bond.

Tertiary: folding which occurs as a result of interactions of the amino acid side chains. These interactions include, but are not limited to the following: covalent disulfide bonds between cysteine residues; interactions of hydrophilic side chains with water; interactions of hydrophobic side chains with each other.

Quaternary: the association of two or more polypeptide chains. Called *oligomeric,* such a protein is made up of more than one *promoter.*

9. There are probably as many different types of proteins as there are different types of structures and functions in living systems. Your text lists the following:

Oxygen transport: hemoglobin, myoglobin

Structural: collagen, keratin, histones

Contractile: actin, myosin

Immune system: immunoglobins

Cross-membrane transport: a variety of proteins in and around membranes, such as receptor proteins.

Regulatory: hormones, perhaps histones

Catalytic: enzymes

10. Enzymes function to regulate catabolic and anabolic activities of cells. They influence (lower) the *energy of activation* thus allowing chemical reactions to occur under conditions which are compatible with living systems. Enzymes possess *active sites* and/or other domains which are sensitive to the environment.

The active site is considered to be a crevice, or pit, which binds reactants, thus enhancing their interaction. The other domains mentioned above may influence the conformation and therefore function of the active site.

11. (a) outcomes from the last cross with its 9:4:3 ratio suggest two gene pairs.

(b) orange = *Y_R_*

yellow = *yyrr, yyR_*

red = *Y_rr*

(c) white ---> yellow --*y*-> red --*r*-> orange

Sample Test Questions (with detailed explanations of answers)

Question 1. The foundations of molecular genetics rest upon the assumption that a genetic material exists with the following properties:

a. Autocatalytic (can replicate itself)

b. Heterocatalytic (can direct form and function)

c. Mutable

d. Can exist in an infinite number of forms

(a) Provide a simple sketch which demonstrates the replication scheme of DNA.

(b) Briefly describe how DNA provides form and function.

(c) At the level of nucleotides, what characterizes mutant DNA?

(d) Why may we say that DNA can exist in an infinite number of forms?

Concepts:

understanding of gene function

understanding DNA structure

function

mutation

variation

Answer 1.

(a) DNA replicates in a semiconservative manner such that each daughter strand is "half-new" and "half-old" in a particular pattern.

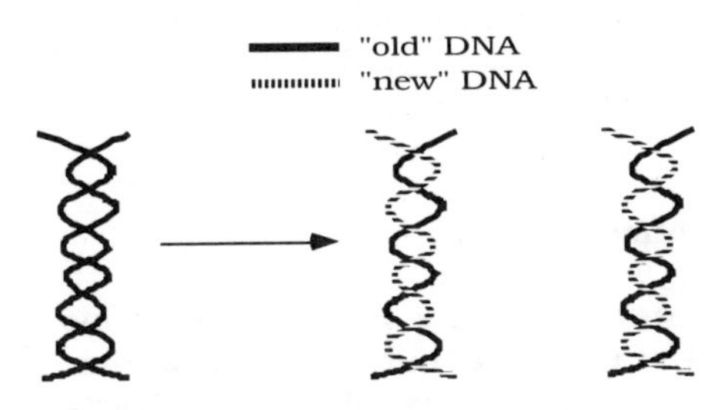

(b) The *Central Dogma of Biology* is based on the production (through transcription) of a RNA messenger from a DNA template and the subsequent "decoding" of that messenger by the process of translation.

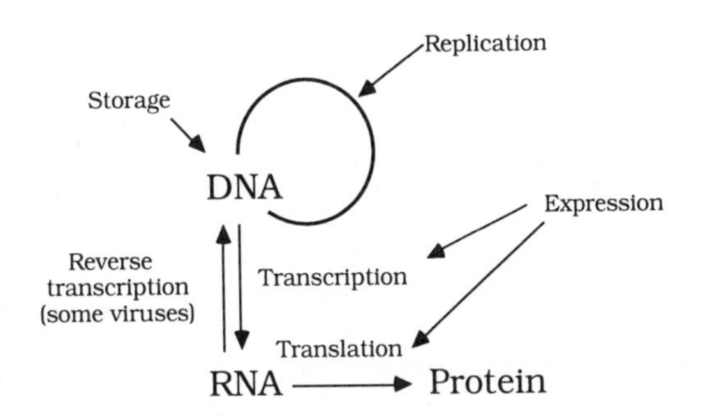

(c) The DNA template contains a sequence of nitrogenous bases which specifies a code from which amino acids are ordered in proteins. Through tautomeric shifts and a number of other natural factors (radiation, chemicals), changes can occur in that sequence of bases. Indeed, the mechanisms by which genes replicate themselves generate errors and leaves us with the conclusion that DNA is an inherently unstable molecule.

(d) Given the variety of organisms and the variation within organisms, there must be numerous, hundreds of millions, elementary factors which are inherited. DNA can provide for this variety by differences in the length and sequence of bases for each inherited functional unit. Given that there are four different types of bases, a sequence having merely ten bases would be capable of 4^{10} (over 1 million) different sequences.

Common errors:

There are usually very few problems with this type of question, except that students often have difficulty clearly explaining that which they know in model form. Written descriptions tend to be more lists of examples rather than explanations of structures and/ or processes.

Question 2. Assume that you are microscopically examining mitotic metaphase cells of an organism with a 2N chromosome number of 2 (both telocentric). Assume also that the cell passed through one S phase labeling (innermost phosphate of dCTP radioactive) just prior to the period of observation.

(a) Draw this cell's chromosomes, and the autoradiographic pattern you would expect to see.

(b)Assuming that the A+T/G+C ratio of the DNA in this cell is 1.67 and this DNA is digested with snake venom diesterase (cleaves at the 3' position), what percentages of the total radioactivity would the following products have?

Adenine_____ Guanine_____

Thymine_____ Cytosine_____

Concepts:

semiconservative replication

chromosome morphology

DNA structure

labeling

enzymatic digestion

5', 3' orientations

Answer 2.

(a) There will be two telocentric metaphase chromosomes in the drawing and each chromatid will be labeled.

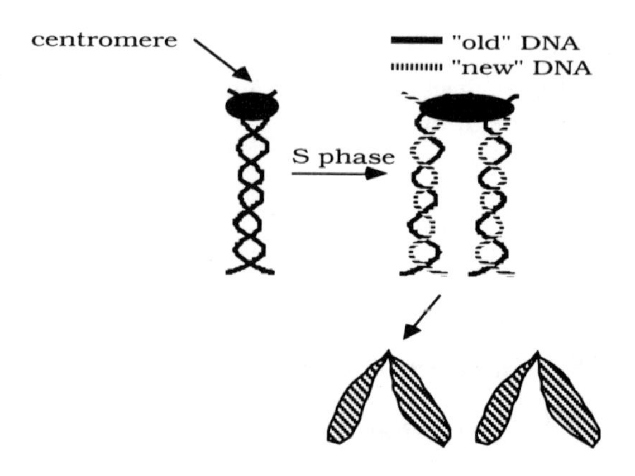

This autoradiographic pattern results because of semiconservative replication. Any cell which contains labeled chromosomes will have had its chromosomes pass through an S phase in the presence of label.

(b) Consider that the DNA was labeled with a dCTP having the innermost phosphate labeled. As this triphosphonucleotide is incorporated into the DNA, it will have the following relationship to its neighbors. *Snake venom diesterase* cleaves DNA at the 3' position meaning that it breaks the bond between the phosphate and the 3' carbon.

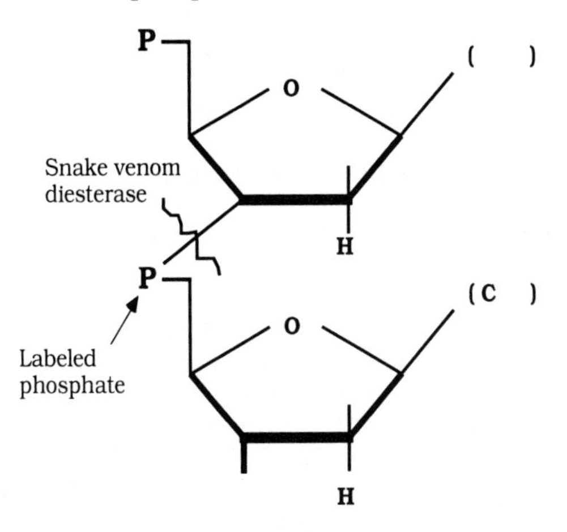

Therefore, the labeled phosphate remains attached to the 5' carbon of the cytosine nucleotide. The A+T/G+C ratio of 1.67 is of no consequence in answering this problem because all of the label remains attached to the cytosine.

Adenine_____ Guanine_____

Thymine_____ Cytosine 100%

Common errors:

inappropriate labeling pattern

inability to draw telocentric chromosomes

understanding cleavage at 3' position

eliminating extraneous information

Question 3. On the graph below draw C_0t curves for DNA from two genomes, one lacking repetitive DNA and the other containing repetitive DNA. Indicate the point on each curve at which the renaturation is half-complete. Label the horizontal and vertical axes accordingly. Explain the molecular basis for the different curves.

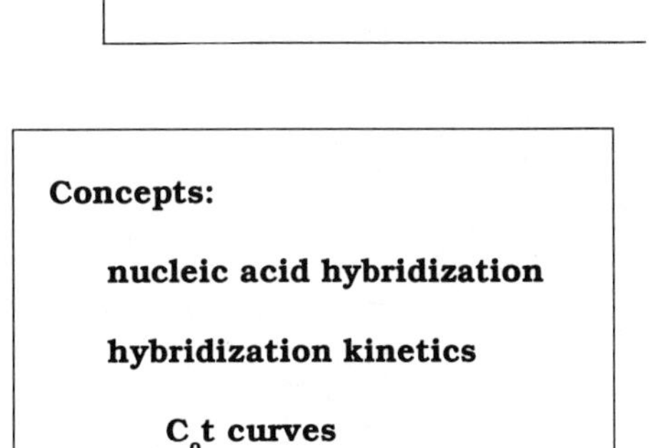

Concepts:

nucleic acid hybridization

hybridization kinetics

C_0t curves

Answer 3. As discussed in the text, the rate of reassociation of melted DNA increases as the proportion of repetitive DNA increases. Such relationships are reflected in C_0t curves in which one plots the fraction of DNA reassociated against the a logarithmic scale of C_0t values which have the units (mole X sec/liter). Because denatured DNA strands which are repetitive have a higher likelihood of complementary interaction, the time of reassociation is less.

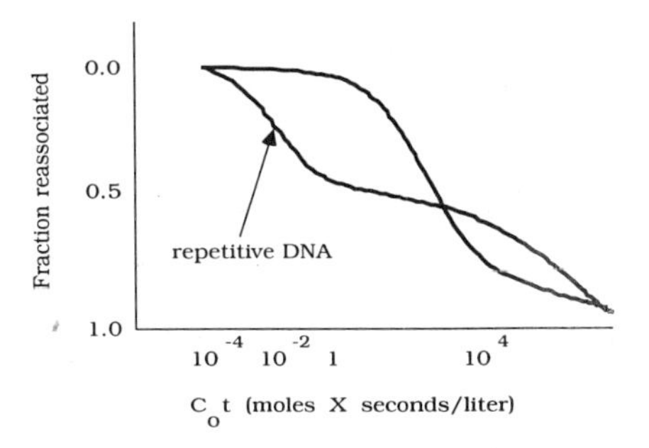

Common errors:

orientation (coordinates) of graph

significance of curve components

repetitive DNA fraction

unique fraction

Question 4. Below is a schematic of transcription and translation occurring simultaneously as described by Miller *et al.* (1970) in *E. coli.* The broken circle symbolizes RNA polymerase. Answer the questions below.

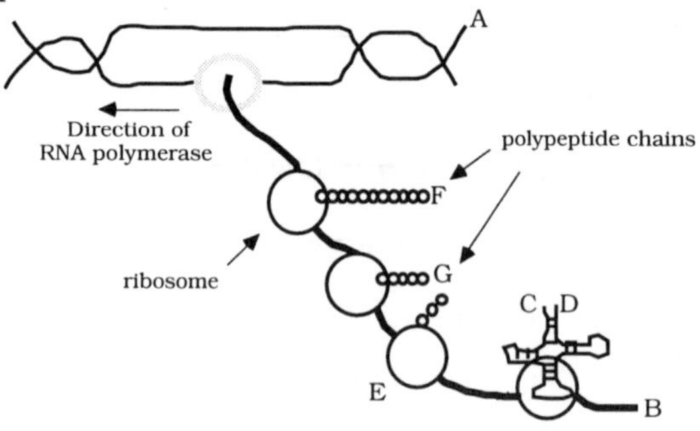

1. Is "A" at the 5' or 3' end of the DNA strand?
______

2. Is "B" at the 5' or 3' end of the RNA strand?
______

3. Is "C" at the 5' or 3' end of the RNA strand?
______

4. What type of RNA is closest to letter "D"?
______

5. Would base sequences near letters "C" or "D" (state which) be expected to hold the amino acid?
_____

6. What is the S value of the rRNA in the small subunit of the ribosome closest to letter "E"?__________

7. Is the amino acid nearest letter "F" the same type as the one nearest letter "G" (yes or no)?
____

Concepts:

transcription

5', 3' orientations

translation

tRNA orientation

general process

rRNA in ribosomes

Answer 4. Overall, this is a drawing of simultaneous transcription and translation in which the RNA polymerase is moving from right to left, making a mRNA which is complementary to one of the two strands of DNA. Ribosomes have added to the nascent (newly forming) mRNA and what appear to be polypeptide chains are protruding from the ribosomes.

(1, 2) In answering this question remember that all synthesis of nucleic acids starts at the 5' end and finishes at the 3' end. Therefore, immediately label the end near point "B" with a 5'. The is the nascent mRNA. Recall that all orientation of complementary strands is antiparallel and since the RNA polymerase is going from right-to-left (according to the arrow) the end of the DNA strand from which the mRNA is copied is the 3' end. Now, since the 3' end of the DNA template strand is identified, its DNA complementary end (at point "A") must be the 5' end.

(3) The codon-anticodon relationship is also antiparallel (based on hydrogen bonding) and since the 5' end of the mRNA is identified, letter "C" must be at the 5' end.

(4) While the diagram is *not to scale*, given the folded structure of the molecule and its position in the ribosome, consider the RNA nearest to letter "D" as tRNA.

(5) Remember that the 3' end of the tRNA holds the amino acid, therefore letter "D" is where the amino acid would be attached.

(6) The tRNA binds mainly to the large subunit of the ribosome, while the mRNA binds mainly to the small subunit of the ribosome. The question asks for the S value of the rRNA in that small subunit. Simultaneous transcription and translation occurs in prokaryotes only (not eukaryotes). The S value for the small subunit of a prokaryotic ribosome is 30S but that value includes both rRNA and protein. The rRNA molecule however has an S value of 16, which is the correct answer.

(7) Since all of the ribosomes are moving along the same mRNA, the amino acid sequences are the same. Therefore, the amino acids nearest the letters "F" and "G" are the same.

Common errors:

polarity of DNA and RNA strands

structure of ribosomes

overall understanding of translation

Question 5. Given below is a single-stranded nucleotide sequence. Answer questions which refer to this sequence.

a. In the circle at the bottom of this sequence, place a 5' or 3', whichever corresponds.

b. Is the above structure an RNA or DNA? State which___________.

c. Assume that a complementary strand is produced in which all the innermost phosphates of the adenine triphosphonucleotide precursors are labeled with ^{32}P. What bases would be labeled if the complementary strand is completely degraded with spleen diesterase (cleaves between the phosphate and the 5'carbon)?________.

d, What bases would be labeled if the complementary strand was completely degraded with snake venom diesterase (cleaves between the phosphate and the 3' carbon)?_________.

Concepts:

DNA and RNA structure

complementarity

5', 3' orientations

degradation products

Answer 5. (a) In this type of drawing, the various carbons of the sugar are readily apparent. It is the orientation and carbon numbering on the sugar which determines the 5'-3' orientation of the molecule. The bottom of the polymer has the 2' and 3' carbons projecting, while the top has the 5' carbon projecting. Therefore the bottom, near the circle is the 3' end of the molecule.

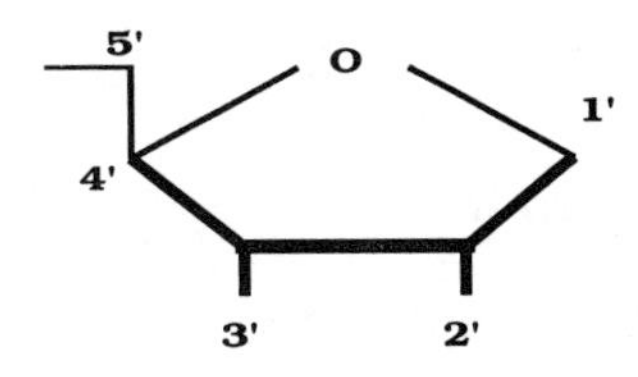

(b) Notice that there is no vertical line protruding from the 2' carbon position in the drawing and that uracil (U) is present. The molecule must therefore be an RNA. **(c)** It is best to start this portion of the problem by roughly drawing the complementary strand, remembering that it will be antiparallel.

If it is a DNA complement, it will have thymine in place of uracil. There is no indication as to the complementary strand being RNA or DNA. Since all ATP's (or dATP's) have at their innermost phosphate a ^{32}P, make certain that they are properly labeled as given below. Since spleen diesterase cleaves between the phosphate and the 5' carbon, the 5' neighbors (C, T or U) will be labeled with the ^{32}P.

(d) Since snake venom diesterase cleaves at the 3' position (between the phosphate and the 3' carbon), the originally labeled ATP (or dATP) will retain the label.

Common errors:

5' to 3' orientations

labeling of complementary strand

understanding enzyme cleavages

Question 6. Assume that you were able to culture a strain of *E. coli* in medium containing either "normal" nitrogen or a heavy isotope of nitrogen (^{15}N). You grow the bacteria for a time in ^{15}N-containing medium which permits one complete replication of the bacterial chromosome. You extract the DNA, calling this extraction A. You continue to grow the bacterial culture in the ^{15}N DNA for a time which permits one more complete round of chromosome replication. You again extract the DNA, calling this extraction B. Assuming that non-labeled DNA has a density of 1.6 and that fully labeled DNA (that is with ***all*** the ^{14}N replaced with ^{15}N) has a density of 1.9, construct sedimentation profiles which reflect the expected densities of DNA from extractions A and B.

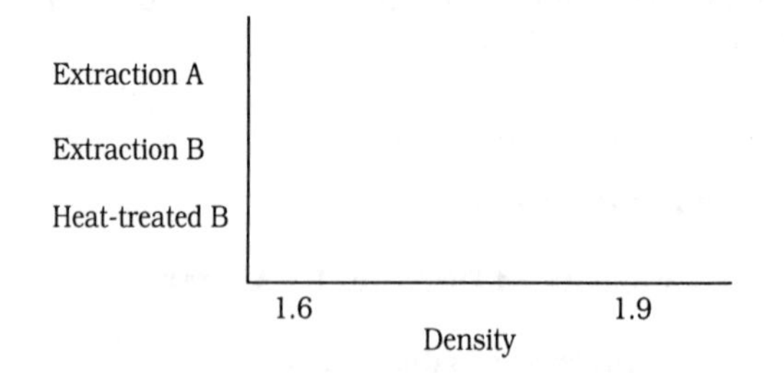

Knowing that heating DNA to 100° C. causes separation of complementary strands, use a broken line (- - -) to indicate the sedimentation profile of heat denaturation of extraction B DNA.

Concepts:

semiconservative replication

labeling

centrifugation

denaturation

Answer 6. Even though circular, in the context of this question, DNA from *E. coli* can be viewed in the following manner. Replication will occur semiconservatively and give the following sedimentation profile.

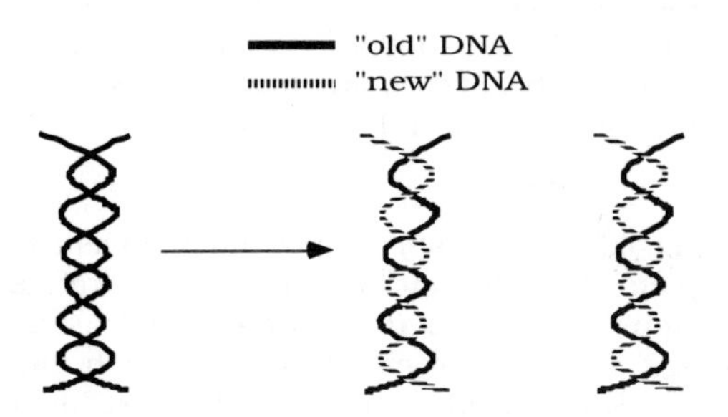

The heat treatment will cause the double-stranded structures to separate, giving the following strands and the profile as shown above.

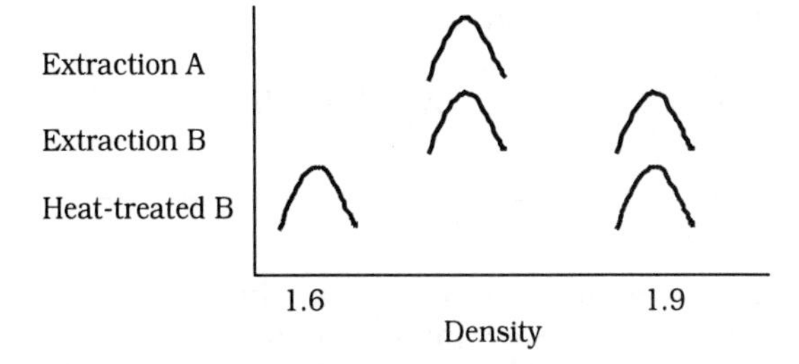

Common errors:

confusion with the labeling experiment

difficulty in seeing sedimentation profiles

application of semiconservative replication

Question 7. Assume that the following sequence of amino acids occurs in a protein starting from the "N" terminus (with asp) of a large polypeptide chain:

asp-glu-ile-leu-ser-thr-met-arg-tyr-try-phe-gly

Assume that gene *X* is responsible for synthesis of this gene. Answer the questions below.

a. Which amino acid(s) would you expect to change if gene *X* is altered by the mutagen, 2-amino purine, such that a transition mutation occurred which caused a change in the 9th base of the mRNA (counting from the 5' end of the coding region)?

b. Which amino acid(s) would you expect to change if gene *X* is altered by the mutagen, acridine orange, such that a frameshift mutation occurred which caused an insertion of a base between bases 3 and 4 of the mRNA (counting from the 5' end of the coding region)?

c. Which amino acid(s) would you expect to change if gene *X* is altered by the mutagen, nitrous acid, such that a mutation occurred which caused a change in the 11th base of the mRNA (counting from the 5' end of the coding region)?

Concepts:

translation

coding

mutation

Answer 7. One of the most frequent difficulties students have with this problem is remembering that the code is triplet and three bases in the mRNA code for each amino acid in a protein. The 5' end of the mRNA corresponds with the N-terminus of the polypeptide chain.

(a) Transition mutations will cause amino acid substitutions. Counting over by "threes" from the N-terminus, the amino acid *ile* should be altered.

(b) Inserting a base between positions 3 and 4 will change the second amino acid; however, recall that acridine orange is a frameshift mutagen and insertion of a base will alter the reading frames for all "downstream" amino acids. Therefore all amino acids in positions two (glu) through twelve will be influenced (excluding degeneracy).

(c) Nitrous acid causes base substitutions, therefore a mutation in the 11th base would influence the amino acid *leu*.

Common errors:

counting amino acids as bases

not understanding what base changes do to amino acid sequences

Question 8. Below is a set of experimental results relating the growth (+) of *Neurospora* on several media. Based on the information provided, present the biochemical pathway and the locations of the metabolic blocks.

Strain	***Medium***		
	MM	*MM+A*	*MM+B*
t409	-	+	+
t410	+	+	+
r3	-	-	+

Concepts:

pathway analysis

Beadle and Tatum "set-up"

biochemical (nutritional) phenotypes

Answer 8. Notice that there are two mutant strains (cannot grow on minimal medium) and one wild type strain (t410). The best way to approach these types of problems, especially when the data are organized in the form given, is to realize that the substance (supplement) which "repairs," as indicated by a (+), a strain is after the metabolic block for that strain. In addition, and most importantly, the substance which "repairs" the highest number of strains either *is the end product* or is *closest to* the end product.

Looking at the table, notice that supplement *B* "repairs" both the mutant strains. Therefore it must be at the end of the pathway or at least after all the metabolic blocks (defined by each mutation). Supplement *A* "repairs" the next highest number of mutant strains (1) therefore it must be second from the end. The pathway therefore would be as follows:

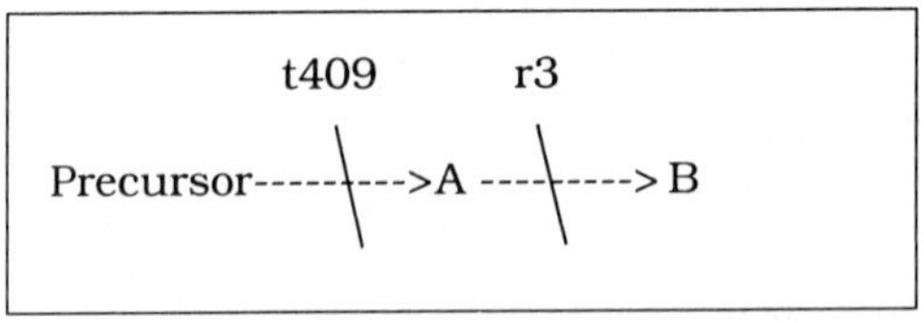

```
               t409         r3
Precursor-----\---->A ----\-----> B
```

To determine the locations at which the strains block the pathway through mutation, apply a similar logic. A block that is "repaired" by all the supplements must be early in the pathway. A block which is "repaired" by only one supplement must be late in the pathway. A supplement which does not "repair" a strain is before that strain's metabolic block.

Common errors:

inability to construct a pathway

failure to see how additives "repair" mutant phenotypes

difficulty in assigning metabolic blocks in pathways

Question 9. Prokaryotes and eukaryotes have evolved different mechanisms for obtaining more than one kind of protein from a single transcription unit (a transcription unit simply being a stretch of DNA that is transcribed into a single primary RNA transcript). Describe two different mechanisms.

> **Concepts:**
>
> **overlapping genes**
>
> **differential hnRNA splicing**

Answer 9. In prokaryotes, overlapping genes present a mechanism for providing two and sometimes more protein products from a single stretch of DNA. In eukaryotes, different sets of introns may be removed, thus providing for a variety of protein products from a single section of DNA. This process is often called *differential splicing*.

> **Common errors:**
>
> **Students often have difficulty in orienting *specific information* they have learned to a general question. If asked about overlapping genes, or differential hnRNA splicing, they would be able to develop an answer. Students sometimes confuse overlapping genes with the non-overlapping code.**

Question 10. Drawn below is a hypothetical protein which contains areas where various types of bonds might be expected to occur. For each area a circle is drawn and in that circle is placed a number. In the corresponding spaces below, state which bond type (or interaction) is most likely illustrated **and** state how that particular type of bond (or interaction) is formed. *You may use a given bond type only once.*

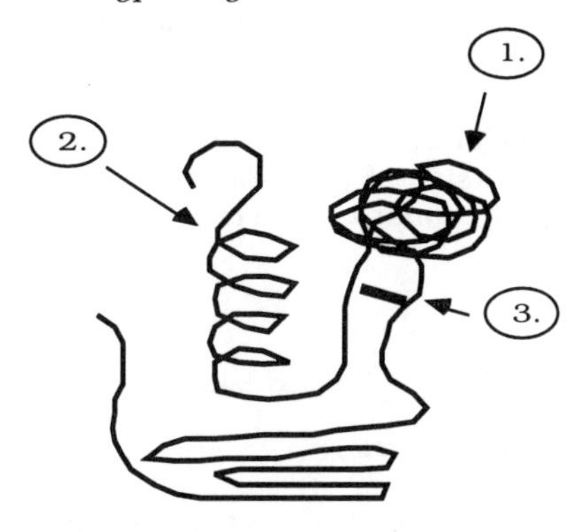

1.
2.
3.
4. What type of amino acids tend to be located on the outside (water side) of the molecule?

> **Concepts:**
>
> **importance of primary structure**
>
> **varieties of bonds**
>
> **"higher level" folding**
>
> **structure/function relationships**

Answer 10.

1. Hydrophobic cluster formed by interaction of hydrophobic amino acids.

2. α helix formed from hydrogen bonds between components of the peptide linkage.

3. Covalent, disulfide bonds formed between cysteine residues

4. The polar amino acids will tend to orient to the outside of the protein where the charged R groups will interact with water.

Common errors:

nature of hydrophobic clustering

understanding of α and β structures

polar side chains and hydrophilic interactions

Question 11. Drawn below is a diagram (not to scale) of DNA in the process of replication. Numbered arrows point to specific structures which you are to identify in the corresponding spaces below:

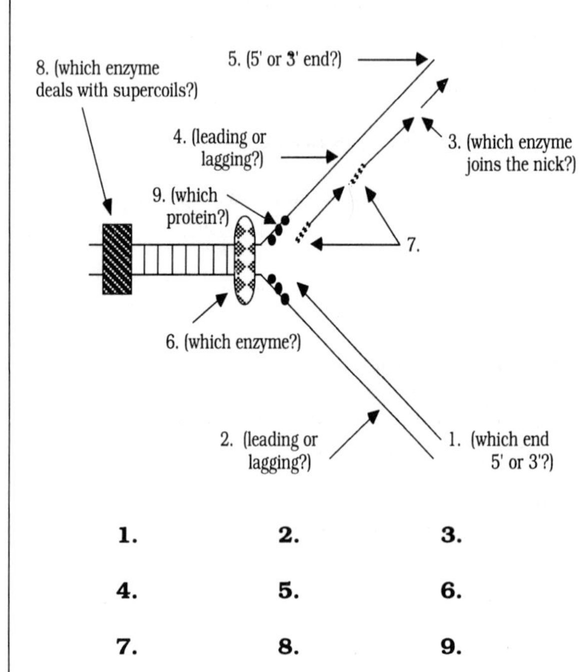

1.	**2.**	**3.**
4.	**5.**	**6.**
7.	**8.**	**9.**

Concepts:

overall DNA replication

5', 3' polarity restrictions

enzymology

priming

Answer 11.

1. This must be the *5' end* of the polymer because all synthesis of polymers is 5' to 3' and the head of the arrow is at the other end of the polymer.

2. The *leading strand* is that strand which is synthesized continuously.

3. A *DNA ligase* will join the nicks.

4. The *lagging strand* is the discontinuous strand.

5. The free end that is complementary to the 5' end at arrow #1 must be the 3'end. That being so, the complement to that 3' end would be 5'. Therefore the *5' end* is at arrow #5.

6. A *helicase* is involved in unwinding the DNA helix.

7. An *RNA primer* is synthesized to initiate DNA synthesis.

8. *DNA gyrase* functions to remove supercoils generated by unwinding the DNA helix.

9. *Single-stranded binding proteins* stabilize the template which is to be replicated.

Common errors:

determination of 5', 3' polarity

naming of enzymes involved

Notes

Notes

Notes

Notes

15

Genetics of Bacteria and Bacteriophages

Vocabulary: Organization and Listing of Terms

Structures and Substances

Minimal medium
- auxotroph
- prototroph
- inoculum

Donor strain, *E. coli* K12
- F sex pilus
- fertility factor, F factor
- *tra*
- Hfr
  - circular chromosome
- F ', merozygotes
  - partial diploid

rec genes, *A, B, C, D*
- DNA - recA protein complex
- D- loop

Plasmid
- F factors
- R plasmids
  - antibiotic resistance

Heteroduplex

Protein coat

Lysozyme

Episome

Prophage P22

Reverse transcriptase

5 - fluorodexyuridine

Temperature sensitive conditional lethals

Processes/Methods

Spontaneous
- fluctuation test
- prototroph
- auxotroph

- lag phase
  - log phase
  - stationary phase
  - serial dilution
- Bacterial recombination
  - conjugation
    - parasexual
    - F^+, F^-
    - physical contact
    - unidirectional
    - F sex pilus
    - donor, "male"
    - recipient, "female"
    - fertility factor, F factor
    - high frequency recombination, Hfr
      - oriented (ordered) transfer
      - interrupted mating technique
      - circular map
      - point of origin
      - low frequency transfer
- F' state
  - merozygotes
  - transformation
    - entry
    - recombination
- cotransformation
  - linkage
- transduction
  - phage life cycle
  - assembly
  - plaque (plaque assay)
    - lysis
    - virulent
  - lysogeny
    - symbiotic relationship
    - prophage
    - temperate phage
    - lysogenic bacterium
    - U-tube experiment
      - filterable agent (FA)
      - prophage P22, etc.
  - generalized transduction (F15.1)
    - abortive transduction
    - complete transduction
- Mutations (viral)
  - *rapid lysis, host range*
  - mixed infection experiments
  - negative interference
  - intergenic exchanges
  - intragenic exchanges

Viral reproduction

- bacteriophage ϕX174
  - (+) strand
  - (-) strand
  - replicative form (RF)
- bacteriophage QB
- poliovirus
- retrovirus
  - reverse transcriptase
  - RNA-directed DNA polymerase
  - oncogenic
  - Rous sarcoma virus (RSV)
  - provirus
  - transformed

Concepts

- Sensitivity and resistance
- Spontaneous mutation
- Genetic variation
- Adaptive mutations (F15.1)
- Bacterial recombination
  - relationship to *rec* genes
- Conjugation
- Transformation
- Transduction
  - lysogeny
- Viral life cycles
  - retroviruses

F15.1. Two forms of *generalized transduction* are depicted below. In *abortive transduction* the entering DNA does not integrate into the bacterial chromosome. In *complete transduction*, the entering DNA integrates into the bacterial chromosome.

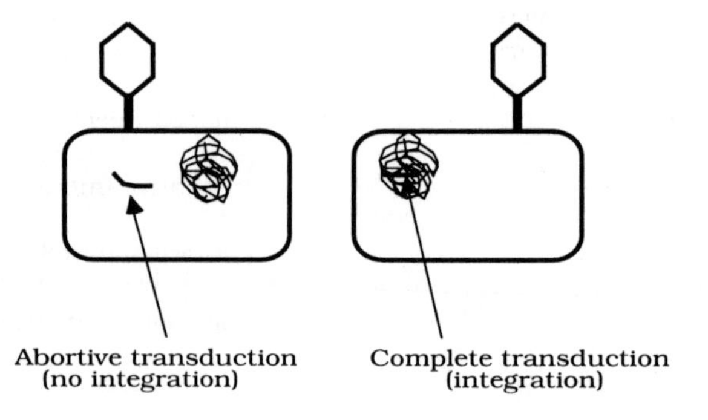

Solutions to Problems and Discussion Questions

1. Three modes of recombination in bacteria are *conjugation*, *transformation*, and *transduction*. Conjugation is dependent on the F factor which, by a variety of mechanisms, can direct genetic exchange between two bacterial cells. Transformation is the uptake of exogenous DNA by cells. Transduction is the exchange of genetic material using a bacteriophage.

2. (a) The requirement for physical contact between bacterial cells during conjugation was established by placing a filter in a U-tube so that the medium can be exchanged but the bacteria can not come in contact. Under this condition, conjugation does not occur.

(b) By treating cells with streptomycin, an antibiotic, it was shown that recombination would not occur if one of the two bacterial strains was inactivated. However, if the other was similarly treated, recombination would occur. Thus, directionality was suggested, with one strain being a donor strain and the other being the recipient.

(c) An F^+ bacterium contains a circular, double-stranded, structurally independent, DNA molecule which can direct recombination. In Hfr cells the F factor is integrated into the bacterial chromosome.

3. (a) In an F^+ X F^- cross, the transfer of the F factor produces a recipient bacterium which is F^+. Any gene may be transferred, and the frequency of transfer is relatively low. Crosses which are Hfr X F^- produce recombinants at a higher frequency than the F^+ X F^- cross. The transfer is oriented (non-random) and the recipient cell remains F^-.

(b) Bacteria which are F^+ possess the F factor, while those that are F^- lack the F factor. In Hfr cells the F factor is integrated into the bacterial chromosome and in F' bacteria, the F factor is free of the bacterial chromosome yet possesses a piece of the bacterial chromosome.

4. Mapping the chromosome in an Hfr X F^- cross takes advantage of the oriented transfer of the bacterial chromosome through the conjugation tube. For each F type, the point of insertion and the direction of transfer are fixed, therefore breaking the conjugation tube at different times produces partial diploids with corresponding portions of the donor chromosome being transferred. The length of the chromosome being transferred is contingent on the duration of conjugation, thus mapping of genes is based on time.

5. One can approach this problem by lining up the data from the various crosses in the following order:

Hfr Strain	*Order*
1	T C H R O >>
2	H R O M B >>
3	<< C H R O M
4	M B A K T >>
5	<< B A K T C

Notice that all of the genes can be linked together to give a consistent map and that the ends overlap, indicating that the map is circular. The order is reversed in two of the crosses indicating the orientation of transfer is reversed.

6. In an Hfr X F^- cross, the F factor is directing the transfer of the donor chromosome. It takes approximately 90 minutes to transfer the entire chromosome. Because the F factor is the last element to be transferred and the conjugation tube is fragile, the likelihood for complete transfer is low.

7. The F^+ element can enter the host bacterial chromosome and upon returning to its independent state, it may pick up a piece of a bacterial chromosome. When combined with a bacterium with a complete chromosome, a partial diploid, or merozygote, is formed.

8. In their experiment a filter was placed between the two auxotrophic strains which would not allow contact. F-mediated conjugation requires contact and without that contact, such conjugation can not occur. The treatment with DNase showed that the filterable agent was not naked DNA.

9. A *plaque* results when bacteria in a "lawn" are infected by a phage and the progeny of the phage destroy (lyse) the bacteria. A somewhat clear region is produced which is called a plaque.

Lysogeny is a complex process whereby certain temperate phage can enter a bacterial cell and instead of following a lytic developmental path, integrate their DNA into the bacterial chromosome. In doing so, the bacterial cell becomes lysogenic. The latent, integrated phage chromosome is called a *prophage.*

10. Starting with a single bacteriophage, one lytic cycle produces 200 progeny phage, three more lytic cycles would produce $(200)^4$ or 1,600,000,000 phage.

11. (a) Culture #1 represents a cell concentration of 240 cells/ml X 10^9 (recall that only 0.1ml of the various dilutions was plated) which had been irradiated. Because leucine is added to the minimal medium, both leu^+ and leu^- cells will grow.

Culture #2 represents a cell concentration of 120 cells/ml X 10^2 which had been irradiated. Because leucine is not added to the minimal medium, only leu^+ cells will grow.

Culture #3 represents a cell concentration of 120 cells/ml X 10^9 which had not been irradiated. Because leucine is added to the minimal medium, both leu^+ and leu^- cells will grow.

Culture #4 represents a cell concentration of 30 cells/ml X 10^1 which had not been irradiated. Because leucine is not added to the minimal medium, only leu^+ cells will grow. One would expect the values to be similar in cultures #1 and #3 because, with leucine added to the medium, one can not differentiate between leu^+ and leu^- cells.

However, it is likely that new non-leucine related nutritional mutations might be induced by the irradiation. If anything, therefore, we might expect to see fewer colonies from culture #1 when compared to #3. The difference of **240 cells/ml** X 10^9 and **120 cells/ml** X 10^9 may be within the limits of experimental error.

(b) The general formula for determining mutation rate is to divide the number of mutant bacteria by the total number of bacteria. In this experiment the spontaneous mutation rate would be calculated as follows:

$$(30 \times 10^1)/(120 \times 10^9) = 0.25 \times 10^{-8}$$

The induced mutation rate would be calculated as follows:

$$(120 \times 10^2)/(240 \times 10^9) = 0.5 \times 10^{-7}$$

12. Reverse transcriptase is required to make a DNA molecule from the RNA genome (template).

13. Notice that the incorporation of loci a^+ and b^+ occurs much more frequently than the incorporation of b^+ and c^+ together (210 to 1) and the incorporation of all three genes $a^+b^+c^+$ occurs relatively infrequently. If a and b loci are close together and both are far from locus c, then fewer crossovers would be required to incorporate the two linked loci compared to all three loci. If all three loci were close together, then the frequency of incorporation of all three would be similar to the frequency of incorporation of any two contiguous loci, which is not the case.

14. The first problem to be solved is the gene order. Clearly, the parental types are

$$a^+b^+c^+ \text{ and } a^-b^-c^-$$

because they are the most frequent. The double crossover types are the least frequent,

$$a^-b^-c^+ \text{ and } a^+b^+c^-.$$

Because it is the gene in the middle that switches places when one compares the parental and double crossover classes, the *c* gene must be in the middle. The map distances are as follows:

a to *c* = (740 + 670 +90 +110)/10,000

= 16.1 map units

c to *b* = (160 + 140 +90 +110)/10,000

= 5 map units

To determine the type of interference, first determine the *expected* frequency of double crossovers (0.161 X .05 = .000805), which when multiplied by 10,000 gives approximately 80. The *observed* number of double crossovers is 90 + 110 or 200. Since many more double crossovers are observed than expected, negative interference is occurring.

15. **(a)** When nutrients A and B are added, selection is for *c*, When nutrients B and C are added, selection is for *a*. When nutrients A and C are added, selection is for *b*.

(b) *b a* c<-------------->c F

16

Recombinant DNA: Cloning and Applications

Vocabulary: Organization and Listing of Terms

Structures and Substances

Restriction endonucleases
- *Eco R1*

Vector
- clone
- cloning vehicle
  - plasmids
    - pUC119
    - pBR322
      - ampicillin, tetracycline
    - polylinker
    - *lacZ*
    - X-gal
  - bacteriophage
  - cosmids
    - phage λ

Cloned DNA fragments

Reverse transcriptase

Ribonuclease H

DNA polymerase I

S_1 nuclease

X-gal

Minisatellite

T cells

Processes/Methods

Recombinant DNA technology
- gene splicing
- genetic engineering
- recombinant DNA
  - restriction endonucleases
    - palindromic sequences
      - sticky ends
      - ligase

Transfection

Library construction

genomic libraries

chromosome-specific libraries

subgenomic fraction

cDNA libraries

reverse transcriptase

ribonuclease H

DNA polymerase I

S_1 nuclease

Replica plating

selection of recombinant clones

probes

colony and plaque hybridization

X-gal

blue, white plaques (colonies)

PCR analysis

denaturation

annealing of specific primers

oligonucleotides

extension of primers

nucleic acid blotting

Southern blot

autoradiography

Northern blot, Western blot

DNA sequencing

applications

gene mapping

RFLPs

neurofibromastosis

multigenerational families

Human Genome Project, HUGO

linkage maps, physical maps

sequencing

disease diagnosis

amniocentesis

chorionic villus sampling

DNA fingerprints

minisatellites

VNTR (variable number tandem repeats)

gene therapy

Severe Combined Immuno-deficiency (SCID)

Adenosine Deaminase (ADA)

commercial applications

protein, antibody production

Concepts

Restriction mapping

Gene mapping

RFLP

sequencing

DNA fingerprinting

Gene therapy, theory and controversy

Solutions to Problems and Discussion Questions

1. Recombinant DNA technology, also called genetic engineering or gene splicing, involves the creation of associations of DNA that are not typically found in nature. Particular enzymes, called *restriction endonucleases*, cut DNA at specific sites and often yield "sticky" ends for additional interaction with DNA molecules cut with the same class of enzyme.

A *vector* may be a plasmid, bacteriophage, or cosmid which receives, through ligation, a piece, or pieces of foreign DNA. The recombinant vector can transform (or transfect) a host cell (bacterium, yeast cell, etc.) and be amplified in number.

2. *Reverse transcriptase* is often used to promote the formation of cDNA (complementary DNA) from a mRNA molecule. Eukaryotic mRNAs typically have a 3' polyA tail as indicated in the diagram below. The poly dT segment provides a double-stranded section which serves to prime the production of the complementary strand.

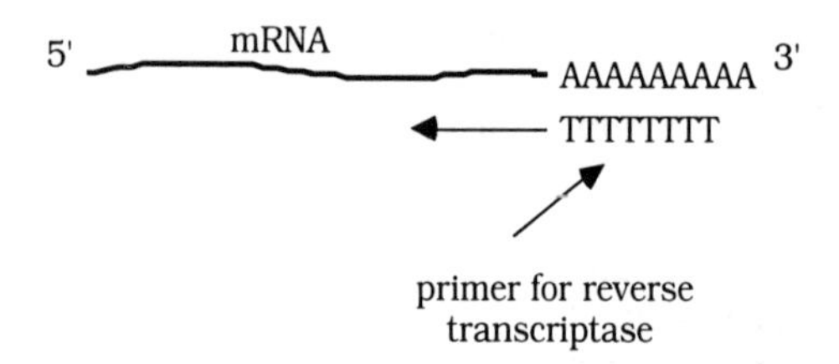

3. (a) Because the *Drosophila* DNA has been cloned into the *Eco R1* site in the ampicillin resistance gene of the plasmid, the gene will be mutated and any bacterium with the plasmid will be ampicillin sensitive. The tetracycline resistance gene remains active however. Bacteria which have been transformed with the recombinant plasmid will be resistant to tetracycline and therefore tetracycline should be added to the medium.

(b) Colonies which grow on a tetracycline medium should be tested for growth on an ampicillin medium either by replica plating or some similar controlled transfer method. Those bacteria which do not grow on the ampicillin medium probably contain the *Drosophila* DNA insert.

(c) Resistance to both antibiotics by a transformed bacterium could be explained in several ways. First, if cleavage with the *Eco R1* was incomplete, then no change in biological properties of the uncut plasmids would be expected. Also, it is possible that the cut ends of the plasmid were ligated together in the original form with no insert.

4. Apply the formula where P is the probability of recovering a given sequence and f represents the fraction of the genome present in each clone.

$$N = \ln(1-P)/\ln(1-f)$$

$$= \ln(1-0.99)/\ln(1-[5 \times 10^3/1.5 \times 10^8])$$

$$= \ln(.01)/\ln(.9999667)$$

$$= -4.605/-0.0000333$$

$$= 1.38 \times 10^5$$

5. When the *insulin* gene is "manufactured," it is made as a complementary copy of the mRNA, which is void of introns. Therefore the insulin mRNA which is made from the cDNA does not have introns.

6. The question of protein/DNA recognition and interaction is a difficult one to answer. Much research has been done to attempt to understand the nature of the specificity of such interactions. In general it is believed that the protein interacts with the major groove of the DNA helix. This information comes from the structure of the few proteins which have been sufficiently well studied to suggest that the DNA major groove and "fingers" or extensions of the protein form the basis of interaction.

7. Given that there is only one site for the action of *Hind*III, then the following will occur. Cuts will be made so that a four base single-stranded set of sticky ends will be produced. For the antibiotic resistance to be present, the ligation will reform the plasmid into its original form. However, two of the plasmids can join to form a dimer as indicated in the diagram below.

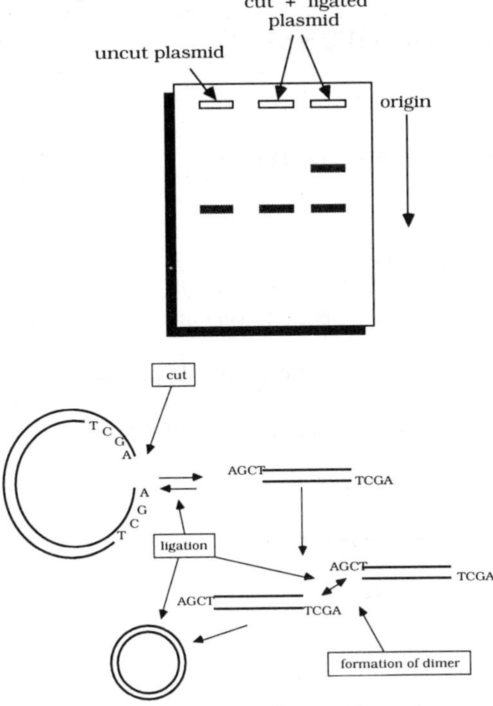

8. The regulated output of a gene, in balance with other gene products in a cell, is necessary for normal physiological function. Regulation is achieved at several levels, from cell-to-cell interactions to position effects caused by neighboring genes. The overall positional, temporal, and molecular environment of a gene, about which there is often little information, will ultimately determine the effectiveness, in a therapeutic sense, of any introduced gene. Before gene therapy is to be dependable, such variables will need to be understood. Even though a great deal is known about the expression of hemoglobin genes, the information presented in the question indicates tissue-specific influences which cause the imbalance leading to thalassemia.

9. Even though there is great economic potential in achieving sophisticated genetic engineering in plants, there are relatively few cloning vectors available for economically important crops.

10. The segment contains the palindromic sequence CCTAGG which is recognized by the restriction enzyme *Bam*HI.

11. Assuming a random distribution of all four bases, the four-base sequence would occur on average every 256 base pairs (4^4), the six-base sequence every 4096 base pairs, and the eight-base sequence every 65,536 base pairs. One might use an eight-base restriction enzyme to produce relatively few large fragments. If one wanted to construct an eukaryotic genome library, such large fragments would have to be cloned into special vectors, such as yeast artificial chromosomes.

12. In the case of haplo-insufficient mutations, gene therapy holds promise; however in "gain-of-function" mutations in all probability, the mutant gene's activity or product must be compromised. Addition of a normal gene probably will not help.

17

Regulation of Gene Expression

Vocabulary: Organization and Listing of Terms

Structures and Substances

Lactose operon
- structural genes
  - *lac Z*
    - β-galactosidase
  - *lac Y*
    - β-galactoside permease
  - *lac A*
    - transacetylase
- polycistronic mRNA
- constitutive mutants
  - *lac I*
  - *lac* O^c
- repressor gene
  - tetramer
  - repressor molecule
  - diffusible, cellular (F17.2)
  - allosteric
    - operator region
      - no diffusible product
      - adjacent control (F17.2)
  - *lac* I^s
- active regulator

Arabinose
- *ara B, A, D*
- *ara* regulatory protein

Tryptophan
- tryptophan synthetase
- *trp* R^-, *trp* R^+
- structural genes
  - *trp E, D, C, B, A*
  - polycistronic mRNA
- *trp P-trp O* region
- leader sequence
  - attenuator

Lambda DNA

- λ repressor protein
  - *cI* repressor
    - O_L, O_R
    - negative control element
    - positive control element
- cro protein
- N, antiterminator

Promoters

- RNA polymerase I, II, III
  - TATA box
  - CAAT box
  - GC box
  - enhancers
  - transcription factors, TFII...
  - upstream activator sequences (UAS)
  - DNA binding domains (structural motifs)
    - zinc fingers
    - helix-turn-helix
    - homeobox
  - homeodomains
    - homeotic genes
    - lucine zippers
  - steroid hormones
    - ecdysone
    - hormone responsive element (HRE)

Processing transcripts

- intron
- exon
  - alternative processing
  - myosin
  - α-crystallin
  - preprotachykinin mRNA
    - tachykinin P, K

Processes/Methods

Genetic regulation

- inducible
  - lactose
- repressible
  - tryptophan
    - attenuation
- negative control (F17.1)
- positive control (F17.1)
- arabinose regulation
- constitutive
- allosteric
- λ lysogeny or lyis

Transcriptional control

- galactose metabolism
- Posttranscriptional regulation
- Posttranslational regulation

Concepts

- Genetic regulation
  - *cis/trans* acting
    - efficiency
      - activity of enzymes
      - level of transcription
  - Positive control (F17.1)
  - Negative control (F17.1)
    - lactose
      - induction
    - tryptophan
      - corepressor
      - repression
      - attenuation
- Operon model
- *Lambda* regulation
- Eukaryotic regulation
  - transcriptional
    - homeosis (homeotic genes)
  - processing
  - posttranscriptional
  - transport
  - translation
- Hormone relationships

F17.1. Illustration of general processes of *negative* and *positive* control. In *negative* control, the regulatory protein inhibits transcription while under *positive* control, transcription is stimulated.

NEGATIVE CONTROL

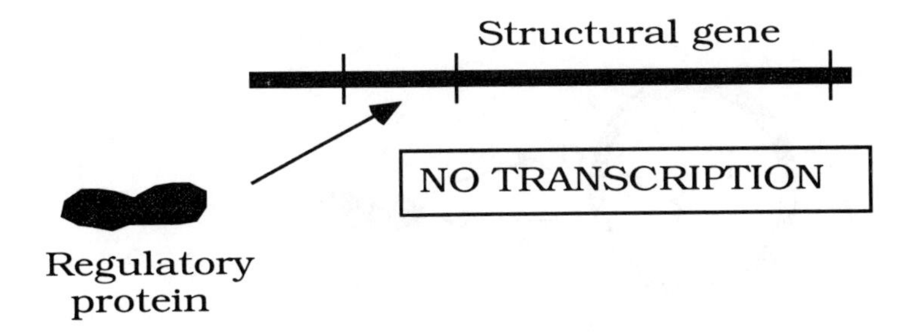

POSITIVE CONTROL

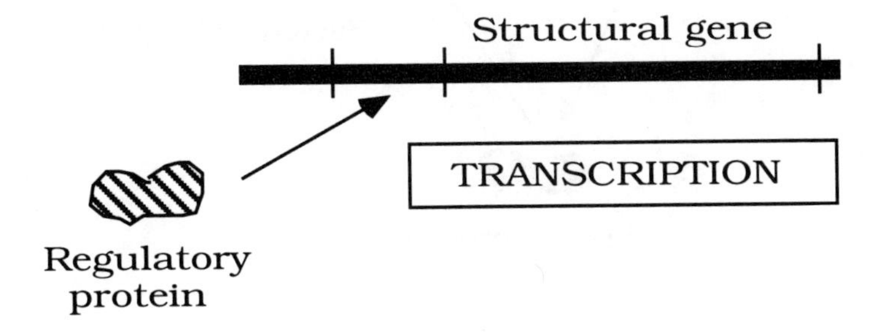

F17.2. Illustration of the nature of the product of the *i* gene. It can act "at a distance" because it is a protein which can diffuse through the cytoplasm and thus act in "trans" as well as in "cis." There is no protein product of the operator gene, therefore it can only act in "cis."

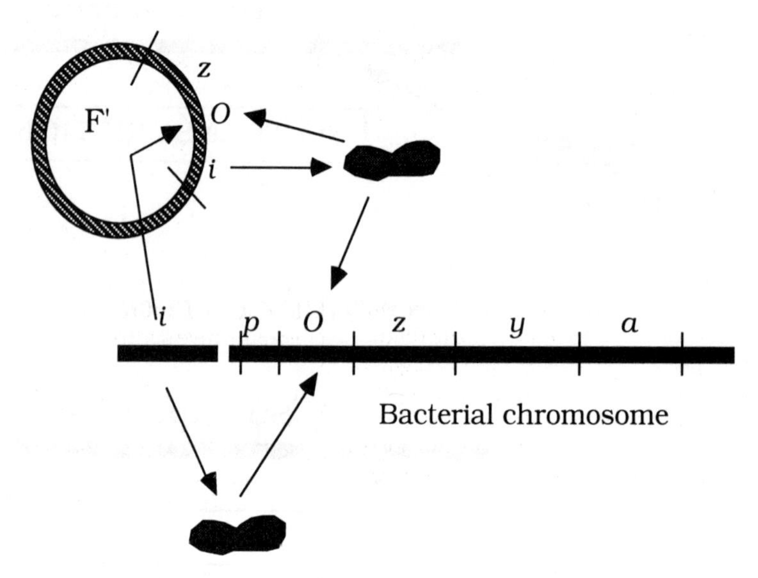

Solutions to Problems and Discussion Questions

1. The answer to this question is a key to enhancing a student's understanding of the Jacob-Monod model as related to lactose and tryptophan metabolism. The enzymes of the lactose operon are needed to break down and use lactose as an energy source. If lactose is the sole carbon source, the enzymes are synthesized *to use* that carbon source. With no lactose present, there is no "need" for the enzymes.

The tryptophan operon contains structural genes for the *synthesis* of tryptophan. If there is little or no tryptophan in the medium, the tryptophan operon is "turned on" to manufacture tryptophan. If tryptophan is abundant in the medium, then there is no "need" for the operon to be manufacturing "tryptophan synthetases."

2. Refer to F17.1 to see that under *negative* control, the regulatory molecule interferes with transcription, while in *positive* control, the regulatory molecule stimulates transcription. Negative control is seen in the *lactose* and *tryptophan* systems as well as a portion of the *arabinose* regulation. Catabolite repression and a portion of the *arabinose* regulatory systems are examples of positive control.

3. In an *inducible system*, the repressor which normally interacts with the operator to inhibit transcription, is inactivated by an *inducer*, thus permitting transcription. In a *repressible system*, a normally inactive repressor is *activated* by a *co-repressor*, thus enabling it (the activated repressor) to bind to the operator to inhibit transcription. Because the interaction of the protein (repressor) has a negative influence on transcription, the systems described here are forms of *negative control* (see F17.1).

4. Refer to the *Essentials* text and to F17.2 to get a good understanding of the lactose system before starting.

$i^+o^+z^+$ = **Inducible** because a repressor protein can interact with the operator to turn off transcription.

$i^-o^+z^+$ = **Constitutive** because the repressor gene is mutant, therefore no repressor protein is available.

$i^+o^cz^+$ = **Constitutive** because even though a repressor protein is made, it can not bind with the mutant operator.

$i^-o^+z^+$ / F' i^+ = **Inducible** because even though there is one mutant repressor gene, the other i^+ gene, on the F factor, produces a normal repressor protein which is diffusible and capable of interacting with the operon to repress transcription. (See F17.2 in this book)

$i^+o^cz^+$ / F' o^+ = **Constitutive** because there is a constitutive operator (o^c) next to a normal z gene. Remembering that this operator functions in *cis* and is not influenced by the repressor protein, constitutive synthesis of β-galactosidase will occur.

$i^so^+z^+$ = **Repressed** because the product of the i^s gene is *insensitive* to the inducer lactose and thus can not be inactivated. The repressor will continually interact with the operator and shut off transcription regardless of the presence or absence of lactose.

$i^so^+z^+$ /F' i^+ = **Repressed** because, as in the previous case, the product of the i^s gene is *insensitive* to the inducer lactose and thus can not be inactivated. The repressor will continually interact with the operator and shut off transcription regardless of the presence or absence of lactose. The fact that there is a normal i^+ gene is of no consequence because once a repressor from i^s binds to an operator, the presence of normal repressor molecules will make no difference.

5. Refer to the *Essentials* text and to F17.2 to get a good understanding of the lactose system before starting.

$i^+ o^+ z^+$ = Because of the function of the active repressor from the i^+ gene, and no lactose to influence its function, there will be **No Enzyme Made.**

$i^+ o^c z^-$ = There will be a **Nonfunctional Enzyme Made** because even though the constitutive operator is in *cis* with a *z* gene, the *z* gene is mutant. The lactose in the medium will have no influence because of the constitutive operator. The repressor can not bind to the mutant operator.

$i^- o^+ z^-$ = There will be a **Nonfunctional Enzyme Made** because with i^- the system is constitutive but the *z* gene is mutant. The absence of lactose in the medium will have no influence because of the non-functional repressor. The mutant repressor can not bind to the operator.

$i^- o^+ z^-$ = There will be a **Nonfunctional Enzyme Made** because with i^- the system is constitutive but the *z* gene is mutant. The lactose in the medium will have no influence because of the non-functional repressor. The mutant repressor can not bind to the operator.

$i^- o^+ z^+ /F' i^+$ = There will be **No Enzyme Made** because in the absence of lactose, the repressor product of the i^+ gene will bind to the operator and inhibit transcription.

$i^+ o^c z^+ /F' o^+$ = Because there is a constitutive operator in *cis* with a normal *z* gene, there will be **Functional Enzyme Made.** The lactose in the medium will have no influence because of the mutant operator.

$i^+ o^+ z^- /F' i^+ o^+ z^+$ = Because there is lactose in the medium, the repressor protein will not bind to the operator and transcription will occur. The presence of a normal *z* gene allows a **Functional and Non-functional Enzyme to be Made**. The repressor protein is diffusable, working in *trans*.

$i^- o^+ z^- /F' i^+ o^+ z^+$ = Because there is no lactose in the medium, the repressor protein (from i^+) will repress the operators and there will be **No Enzyme Made**.

$i^s\ o^+\ z^+ / F' o^+$ = With the product of i^s there is binding of the repressor to the operator and therefore **No Enzyme Made.** The lack of lactose in the medium is of no consequence because the mutant repressor is insensitive to lactose.

$i^+ o^c z^- /F' o^+ z^+$ = The arrangement of the constitutive operator (o^c) with the mutant *z* gene will cause a **Nonfunctional Enzyme to be Made.** In addition, the normal repressor will be inactivated by the lactose and the unit residing in the F factor will produce a **Functional Enzyme to be Made.** See F17.2.

6. First notice that in the first row of data, the presence of <u>tm</u> in the medium causes the production of active enzyme from the wild type arrangement of genes. From this one would conclude that the system is *inducible*. To determine which gene is the structural gene, look for the IE function and see that it is related to c^-. Therefore *c* codes for the **structural gene.** Because when *b* is mutant, no enzyme is produced, *b* must be the **promoter**.

Notice that when genes *a* and *d* are mutant, constitutive synthesis occurs, therefore one must be the operator and the other gene codes for the repressor protein. To distinguish these functions, one must remember that the repressor operates as a diffusible substance and can be on the host chromosome or the F factor (functioning in *trans*). However, the operator can only operate in *cis*. In addition, in *cis*, the constitutive operator is dominant to its wild type allele, while the mutant repressor is recessive to its wild type allele.

Notice that the mutant *a* gene is dominant to its wild type allele, whereas the mutant *d* allele is recessive (behaving as wild type in the first row). Therefore, the *a* locus is the **operator** and the *d* locus is the **repressor** gene.

7. The *cI* gene is responsible for repressing the genes which control the lytic cycle of λ phage. If the *cI* gene is mutant then the lysogenic cycle could not occur. Interestingly, there are temperature-sensitive alleles of the *cI* gene which maintain the lysogenic state at one temperature, but cause the lytic cycle at an elevated temperature.

8. There are several reasons for anticipating a variety of different regulatory mechanisms in eukaryotes as compared to prokaryotes. Eukaryotic cells contain greater amounts of DNA and this DNA is associated with various proteins, including histones and nonhistone chromosomal proteins. *Chromatin* as such does not exist in prokaryotes. In addition, whereas there is usually only one chromosome in prokaryotes, eukaryotes have more than one chromosome all enclosed in a membrane (nuclear membrane). This nuclear membrane separates, both temporally and spatially, the processes of transcription and translation, thus providing an opportunity for post-transcriptional, pre-translational regulation.

While prokaryotes respond genetically to changes in their external environment, cells of multicellular eukaryotes interact with each other as well as the external environment. The structural and functional diversity of cells of a multicellular eukaryote, coupled with the finding that all cells of an organism contain a complete complement of genes, suggests that in some cells certain genes are active which are not active in other cells.

It is often difficult to study eukaryotic gene regulation because of the complexities mentioned above, especially tissue specificity and the various levels at which regulation can occur. Obtaining a homogeneous group of cells from a multicellular organism often requires a significant alteration of the natural environment of the cell. Thus, results from studies on isolated cells must be interpreted with caution. In addition, because of the variety of intracellular components (nuclear and cytoplasmic) it is difficult to isolate, free of contamination, certain molecular species. Even if such isolation is accomplished, it is difficult to interpret the actual behavior of such molecules in an artificial environment.

9. ***Organization of DNA:*** Changes in DNA/chromosome structure can influence overall gene output.

Gene amplification refers to cases where an increase in gene products is achieved by an increase in the number of genes producing those products. Such amplification can be achieved intrachromosomally (chorion genes) or extrachromosomally (rRNA genes in some amphibians).

Gene rearrangement involves the recombination of gene segments, which changes the types and amounts of gene products (antibody variability).

Transposable elements may alter gene output by influencing promoters or enhancers or by introducing changes which alter gene coding or termination signals. It is also possible that transposable elements introduce novel promoters and/or enhancers.

Transcription: There are several factors which are known to influence transcription: *promoters*, TATA, CAAT, and GC boxes, as well as other upstream regulatory sequences; *enhancers*, which are *cis*-acting sequences that act at various locations and orientations; *transcription factors*, with various structural motifs (zinc fingers, homeodomains, and leucine zippers) which bind DNA and influence transcription; *receptor-hormone complexes* which influence transcription.

Processing and transport types of regulation involve the efficiency of hnRNA maturation as related to capping, polyA tail addition, and intron removal.

Translation: After mRNAs are produced from the processing of hnRNA, they have the potential of being translated. The stability of the mRNAs appears to be an additional regulatory control point. Certain factors, such as protein subunits may influence a variety of steps in the translational mechanism. For instance, a protein or protein subunit may activate an RNAse which will degrade certain mRNAs or a particular regulatory element may cause a ribosome to stall, thus decreasing the speed of translation and increasing the exposure of a mRNA to the action of RNAses.

10. *Promoters* are conserved DNA sequences which influence transcription from the "upstream" side (5') of mRNA coding genes. They are usually fixed in position and within 100 base pairs of the initiation site for mRNA synthesis. Examples of such promoters are the following: TATA, CAAT, and GC boxes.

Enhancers are *cis*-acting sequences of DNA which stimulate the transcription from most, if not all, promoters. They are somewhat different from promoters in that the position of the enhancer need not be fixed; it may be upstream, downstream, or within the gene being regulated. The orientation may be inverted without significantly influencing its action. Enhancers can work on different genes, that is, they are not gene-specific.

11. Because the deletion of the regulatory gene causes a loss of synthesis of the enzymes, the regulatory gene product can be viewed as one exerting *positive control.* When tis is present, no enzymes are made, therefore, tis must inactivate the positive regulatory protein. When tis is absent, the regulatory protein is free to exert its positive influence on transcription. Mutations in the operator negate the positive action of the regulator. A model which illustrates these points is presented in the next column.

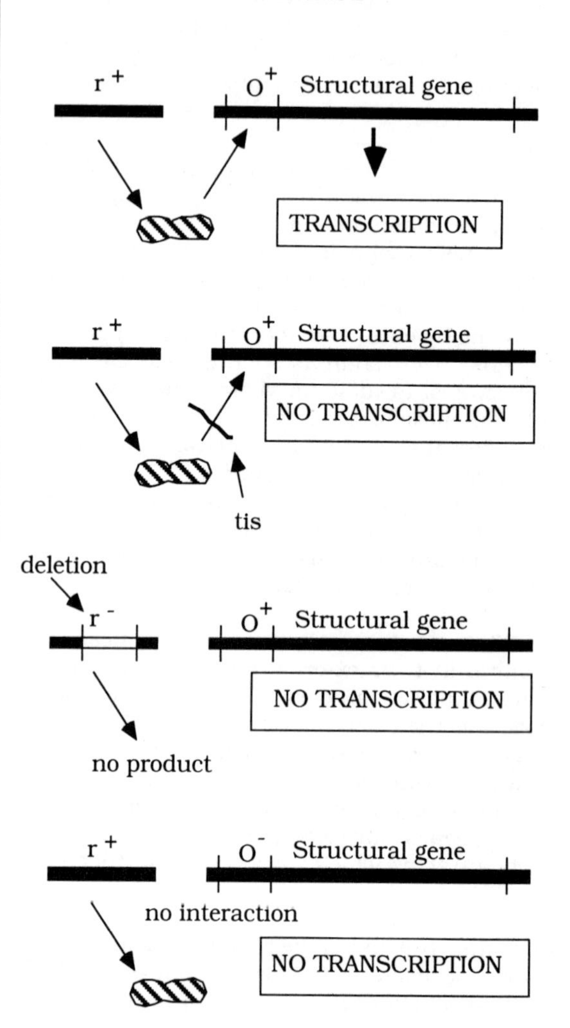

18

Developmental Genetics and Behavior

Vocabulary: organization and listing of terms

Structures and Substances

- *Rana*
- *Xenopus*
- *Drosophila*
- Molecular gradients
  - anterior-posterior axis
    - *bicoid*
  - maternal genes
- Maternal cytoplasm
- Cortex
- Blastoderm
- Imaginal disks
- Ethylmethanesulfonate

Processes/Methods

- Development
  - determination
  - differentiation
    - variable gene activity
    - genetic and morphological changes
  - cell-cell interaction
  - cellular specialization
  - morphogenesis
  - extracellular environment
    - maternal cytoplasm
- Analyses
  - newt embryo (Spemann)
  - *Rana pipiens* (Briggs and King)
  - *Xenopus laevis* (Gurdon)
    - enucleation of oocytes
    - serial transfers
    - gene reactivation
  - *Drosophila*
    - embryogenesis
    - blastoderm

imaginal disks

fate maps

compartments

segmentation genes

gap

pair rule

segment polarity

metamorphosis

molecular gradients

anterior-posterior

bicoid, nanos, etc.

dorsal-ventral

selector genes

homeotic mutants

bithorax complex

antennapedia complex

homeobox

homeodomain

positional cues

Behavior

taxis

chemotaxis

thermotaxis

geotaxis

phototaxis

movement patterns

genetic/mosaic dissection (*Drosophila*)

attached-X

electroretinogram

focus (F18.3)

ring-X chromosome (F18.3)

blastoderm

Learning in *Drosophila*

learning/memory mutants

Molecular biology of behavior

Human behavior genetics

Huntington disease

Lesch-Nyhan syndrome

manic depression

schizophrenia

Concepts

Development (F18.1)

determination

differentiation

cell-cell interaction

Differential gene action

Developmental genetics

molecular explanations

Variable gene activity (F18.2)

differential transcription

Stability of differentiation

transdetermination

Genomic equivalence

Totipotent

Maternal influences

anterior-posterior gradient

positional information

segmentation genes

Homeotic genes

homology

Cellular/extracellular environment

Nature/nurture controversy

Focus of a gene (F18.3)

F18.1. Illustration of the relationship between *determination* and *differentiation*. Determination sets the program which will later be revealed during differentiation.

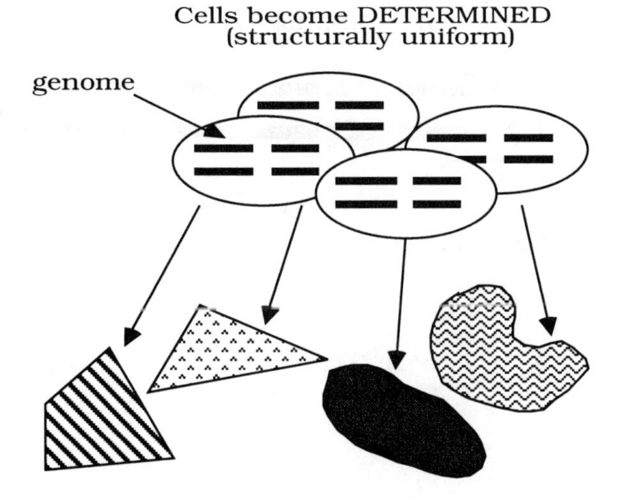

F18.2. Illustration of the genomic changes which are thought to occur during cell differentiation. The *variable gene activity model* states that different sets of genes are transcriptionally active in differentiated cells.

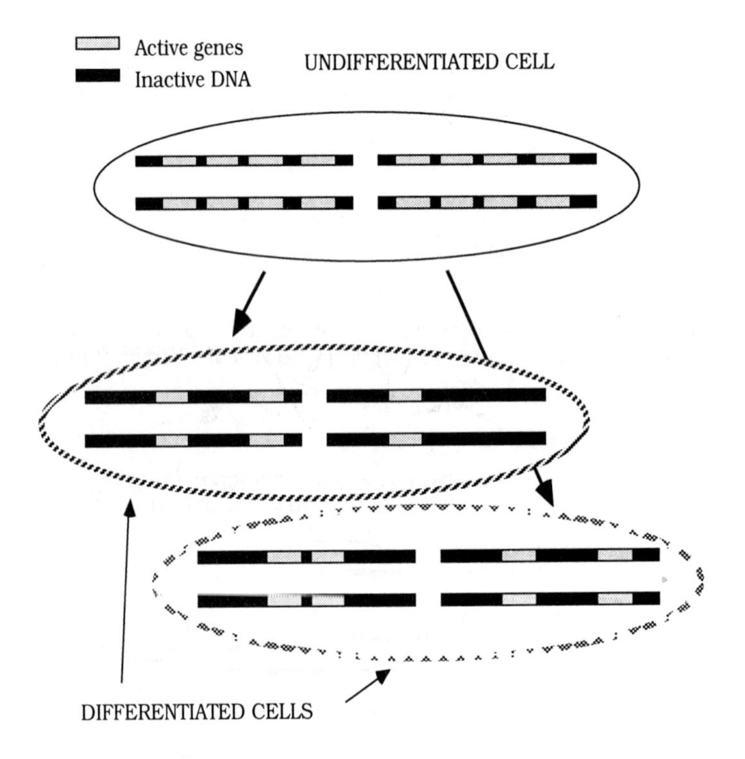

F18.3. Illustration of the procedure of using the ring-X chromosome and appropriate markers (*singed bristles*) to generate gyndromorphic flies for mapping the primary focus of a gene (*limp, lp*). After elimination of the unstable ring-X chromosome in one of the daughter cells at the first mitotic division, male (XO) and female (XX) tissues are produced in the same fly. Having the *singed bristles* gene in the heterozygous state allows one to identify the male and female tissues. Male tissue will show the singed bristles phenotype.

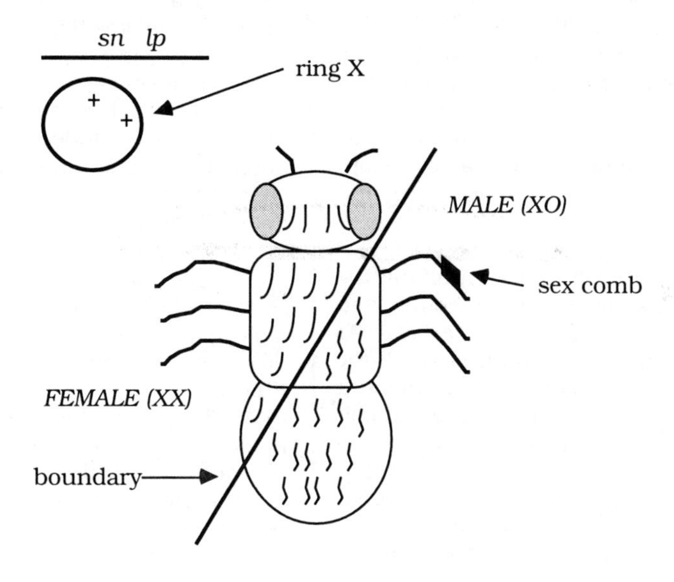

Solutions to Problems and Discussion Questions

1. *Determination* refers to early developmental and regulatory events which set eventual patterns of gene activity. Determination is not the end result of the regulatory activity, rather, it is the process by which the genomic fate of a particular cell type is fixed. *Differentiation* on the other hand follows determination and is the manifestation, in terms of genetic, physiological, and morphological changes, of the determined state.

2. Many of the appendages of the head, including the mouth parts and the antennae, are evolutionary derivatives of ancestral leg structures. In *spineless aristapedia,* the distal portion of the antenna is replaced by its ancestral counterpart, the distal portion of the leg (tarsal segments). Because the replacement of the arista (end of the antenna) can occur by a mutation in a single gene, one would consider that one "selector" gene distinguishes aristal from tarsal structures. Some support for this hypothesis comes from the pattern of transdetermination, where a "one-step" change is involved in the interchange of leg and antennal structures.

3. Actinomycin D is useful in determining the involvement of transcription on molecular and developmental processes. Because maternal RNAs are present in the fertilized sea urchin egg (as is the case with many egg types) a considerable amount of development can occur without transcription. Because gastrulation is inhibited by *prior* treatment with actinomycin D, it would appear that earlier gene products are necessary for the initiation and/or continuation of gastrulation. Clearly a critical period (6th to 11th hours of development) exists for gastrulation in which gene activity is required.

4. The fact that nuclei from almost any source remain transcriptionally and translationally active substantiates the fact that the genetic code and the ancillary processes of transcription and translation are compatible throughout the animal kingdom.

Because the egg represents an isolated, "closed" system which can be mechanically, environmentally, and to some extent biochemically manipulated, various conditions may be developed which allow one to study facets of gene regulation. For instance, the influence of transcriptional enhancers and suppressors may be studied along with factors which impact on translational and post-translational processes. Combinations of injected nuclei may reveal nuclear-nuclear interactions which could not normally be studied by other methods.

5. The egg is not an unorganized collection of molecules from which life springs after fertilization. It is a highly organized structure, "preformed" in the sense that maternal informational molecules are oriented to provide an anterior-posterior and dorsal-ventral pattern from which nuclei receive positional cues.

Such positional cues lead to the "determined" state, from which cells later reveal their adult form (differentiation). Indeed, in *Drosophila* and many other organisms, embryonic fate maps may be constructed, thereby attesting to the maternally-derived "prepattern" present in the egg.

The egg therefore *is* preformed, not in the sense that a miniature individual resides, but in a molecular prepattern upon which development depends. However, work by Spemann, Briggs and King, and Gurdon, indicates that there is plasticity in the programming of nuclei and that even nuclei from somewhat specialized cells often have the potential to direct the development of the entire adult individual. Such *totipotent* behavior of cells indicates that development arises as a result of a series of progressive steps in which cells acquire new structures and functions as development progresses. The epigenetic theory is viewed today as the result of differential gene expression.

6. The advantages of *Drosophila* in behavioral studies include the following. First, *Drosophila* has an immense repertoire of chromosomal and single-gene alterations, which allows the investigator the opportunity to create a broad range of genetic circumstances which facilitate the isolation and characterization of behavioral mutants. A good example is that of generating gynandromorphs (mosaics) for identifying the primary focus of a gene that determines a particular behavior (see F18.3). Second, *Drosophila* has a fairly elaborate set of various behaviors (reproductive, locomotor, taxic, etc.) which allow one to generate and isolate a variety of abnormal behaviors upon which genetic studies depend.

7. One of the easiest ways to determine whether a genetic basis exists for a given abnormality is to cross the abnormal fly to a normal fly. If the trait is determined by a dominant gene, the trait should appear in the offspring, probably half of them if the gene was in the heterozygous state. If the gene is recessive and homozygous, then one may not see expression in the offspring of the first cross, however if one crosses the F_1 one should see the trait appearing in approximately 1/4 of the offspring. Modifications of these patterns would be expected if the mode of inheritance is X-linked or shows other modifications of typical Mendelian ratios.

One might hypothesize that the focus is in the nervous or muscular system. Mapping the primary focus of the gene could be accomplished with patience and the use of the unstable ring-X chromosome to generate gynandromorphs. Given that the gene is X-linked, one would use classical recombination methods to place a recessive X-linked marker, such as *singed bristles,* on the X chromosome with the gene causing the limp. This would help one identify the male/female boundaries. One would then cross homozygous (for the trait and markers) females to ring-X males, or the reciprocal, then examine the offspring for gynandromorphs (and singed mosaics).

If one obtained a pool of gynandromorphs, one could then assess the phenotype (limp or normal) with respect to exposure of the recessive gene in the male tissue. Correlating such would allow one to provide an educated guess as to the primary focus of the gene causing the limp. See F18.3.

8. From the information provided, one can conclude that the tasting trait is determined by a dominant gene. Notice that a 3:1 ratio of tasters to nontasters is obtained in the third cross. This result strongly argues for the *TT* or *Tt* condition providing taste of PTC. Information provided by the other crosses reinforces this hypothesis.

9. Several problems in the study of human behavioral genetics would include the following.

1. With a relatively small number of offspring produced per mating, standard genetic methods of analysis are difficult.

2. Records on family illnesses, especially behavioral illnesses, are difficult to obtain.

3. The long generation time makes longitudinal (transmission genetics) studies difficult.

4. The scientist cannot direct matings that will provide the most informative results.

5. The scientists can't always subject humans to the same types of experimental treatments as other organisms.

6. Traits that are of interest to study are often extremely complex and difficult to quantify.

19

Genetics and Cancer

Vocabulary: organization and listing of terms

Structures and Substances

- *Saccharomyces cerevisiae*
- *Schizosaccharomyces pombe*
  - kinases, CDK
  - cyclins
- Tumor suppressor genes
  - proto-oncogenes (*c-onc*)
  - oncogenes
  - *v-onc*
  - other oncogenes
- EF2
  - zinc finger domains
- Sarcoma
- Adenoma
- Metastatic tumor
  - extracellular matrix
  - metalloproteinases
  - inhibition (TIMP)
- Retrovirus
  - Rous sarcoma virus
  - reverse transcriptase
    - acute transforming virus
    - nonacute (nondefective) virus
- p53, guardian of the genome
  - apoptosis
- Hepatitis B virus (HBV)
  - hepatocellular carcinoma

Processes/Methods

- Cancer
  - somatic vs. germ-line
  - chromosomal changes
    - leukemia (Down syndrome)
    - chronic myeloid leukemia

- Philadelphia chromosome
  - hybrid genes
  - lymphoma
- environmental factors
  - ionizing radiation
  - chemicals
  - sunlight
  - epidemiological studies
  - drugs
  - familial adenomatous polyposis (FAP)

Cell cycle control
- G_1, G_0, S, G_2
- "start"
- S phase control point

Genetic predisposition
- familial adenomatous polyposis (FAP)
  - multistep, multigenic
- retinoblastoma, chromosome 13
  - 90% "penetrant"
  - familial
  - sporatic
- phosphorylation
- dephosphorylation
- EF2
- general regulator

Wilms tumor, chromosome 11
- autosomal dominant
- tissue specific regulator

LiFraumeni syndrome
- autosomal dominant

Tumor suppressor genes

Proto-oncogenes

Oncogenes
- point mutations (*ras*)
- translocations (*c-abl*)
- overexpression
  - promoters, enhancers
  - amplification

metastasis

colon cancer
- genetic factors
- *p53*

Concepts

Cellular basis of cancer
- cell cycle control

Genetic influences
- suppressor genes
- protooncogenes
- oncogenes

Model for retinoblastoma control

Model for Wilms tumor control

Origin of oncogenes

Predisposition

Solutions to Problems and Discussion Questions

1. Familial retinoblastoma is inherited as an autosomal dominant gene with 90% penetrance, that is, 90% of the individuals which inherit the gene will develop eye tumors. The gene usually expresses itself in youngsters. Because the husband's sister has RB, one of the husband's parents has the gene for RB and the husband has a 50:50 chance of inheriting that gene. However, because the husband is past the usual age of onset, it is quite likely that he was lucky and did not receive the RB gene. In that case, the chance of a child born to this couple having RB is no higher than the frequency of sporatic occurrence. However, because the gene is 90% penetrant, there is a chance that the husband has the gene but does not express it. The probability of that occurrence would be 0.50 (of inheriting the gene) X 0.10 (not expressing the gene) = 0.05. The chance of the husband then passing this non-expressed gene to his child would be again 0.5, so 0.50 X 0.05 = 0.025 for the child inheriting this gene. If the child inherits the RB gene, he/she has a 90% chance of expressing it. Therefore the overall probability of the child having RB (using this logic) would be 0.025 X 0.9 = 0.0225 or just over 2% (or just over 1 in 50).

To test the presence of the RB gene in the husband, it is possible in some forms of RB to identify (by Southern blot) a defective or missing DNA segment. Otherwise, one might attempt to assay the RB product in cells to see if it is present and functional at normal levels.

2. Review Chapter 2 of the *Essentials* text and note that the following stages of the cell cycle are discussed: G_1, G_0, S, G_2. The G_1 stage begins after mitosis and is involved in the synthesis of many cytoplasmic elements. In the S phase DNA synthesis occurs. G_2 is a period of growth and preparation for mitosis. Most cell cycle time variation is caused by changes in the duration of G_1. G_0 is the non-dividing state.

3. The major regulatory points of the cell cycle include the following:

1. "Start" in late G1

2. A point in early S phase

3. The border between G2 and mitosis

4. Kinases regulate other proteins by adding phosphate groups. Cyclins bind to the kinases, switching them on and off. In the S phase, cyclin A combines with a kinase to regulate the initiation of DNA synthesis. At the G_2/mitosis border a kinase combines with another cyclin (cyclin B).

5. To say that a particular trait is inherited conveys the assumption that when a particular genetic circumstance is present, it will be revealed in the phenotype. For instance, albinism is inherited in such a way that individuals who are homozygous recessive express albinism. When one discusses an inherited predisposition, one usually refers to situations where a particular phenotype is expressed in families in some consistent pattern; however, the phenotype may not always be expressed or may manifest itself in different ways. In retinoblastoma, the gene is inherited as an autosomal dominant and those that inherit the mutant RB allele are predisposed to develop eye tumors. However, approximately 10% of the people known to inherit the gene don't actually express it and in some cases expression involves only one eye rather than two.

6. A tumor suppressor gene is a gene that normally functions to suppress cell division. Since tumors and cancers represent a significant threat to survival and therefore Darwinian fitness, strong evolutionary forces would favor a variety of co-evolved and perhaps complex conditions in which mutations in these suppressor genes would be recessive. Looking at it in another way, if a tumor suppressor gene makes a product that regulates the cell cycle favorably, cellular conditions have evolved in such a way that sufficient quantities of this gene product are made from just one gene (of the two present in each diploid individual) to provide normal function.

7. The dominantly inherited RB gene, located on chromosome 13, encodes a 928 amino acid protein which is present in all cell and tissue types in G0 cells and those active in the cell cycle. When the RB protein is dephosphorylated, it acts to suppress cell division by binding to and inactivating a transcription factor (EF2). Wilms tumor is caused by an autosomal dominant gene which also functions as a tumor suppressor. The WT gene is located on the short arm of chromosome 11 and encodes a protein with four contiguous zinc finger domains, which are characteristic DNA binding proteins. The gene is activated only in mesenchymal cells of the fetal kidney and in the tumorous nephroblastoma cells. It is suggested that the WT gene product acts directly to turn off genes that sustain cell proliferation or turn on genes which differentiate mesenchymal cells into kidney cells. Both RB and WT gene products are restricted to the nucleus but the RB protein does not bind to DNA; rather, it appears to be a general regulator of cell division while the WT gene product is a cell or tissue-specific regulator.

8. Oncogenes are genes that induce or maintain uncontrolled cellular proliferation associated with cancer. They are mutant forms of proto-oncogenes which normally function to regulate cell division.

9. A translocation involving exchange of genetic material between chromosomes 9 and 22 is responsible for the generation of the "Philadelphia chromosome." Genetic mapping established that certain oncogenes were combined to form a hybrid gene that encodes a 200kd protein which has been implicated in the formation of chronic myelocytic leukemia.

10. Unfortunately, it is common to spend enormous amounts of money on dealing with diseases after they occur rather than concentrating on disease prevention. Too often pressure from special interest groups or lack of political stimulus retards advances in education and prevention. Obviously, it is less expensive, both in terms of human suffering and money, to seek preventive measures for as many diseases as possible. However, having gained some understanding of the mechanisms of disease, in this case cancer, it must also be stated that no matter what preventive measures are taken, it will be impossible to completely eliminate disease from the human population. It is extremely important, however, that we increase efforts to educate and protect the human population from as many hazardous environmental agents as possible.

11. Any agent which causes damage to DNA is a potential carcinogen since cell cycle control is achieved by gene (DNA) products, known as proteins. Since cigarette smoke is known to contain an agent which changes DNA, in this case transversions, numerous modified gene products (including cell cycle controlling proteins) are likely to be produced. The fact that many cancer patients have such transversions in *p53* strongly suggests that cancer is caused by agents in cigarette smoke.

20

The Genetic Basis of Immunity

Vocabulary: Organization and Listing of Terms

Structures and Substances

- Immune system
  - antibodies
    - B cells
    - T cells
      - helper cells
      - suppressor cells
      - killer cells
      - T4 helper/inducer cells
      - T8 cytotoxic/suppressor cells
  - antigens
  - phagocyte
    - antigen-presenting cells
    - macrophage
  - memory cell
  - plasma cell
  - perforin
- immunoglobins
  - IgG, IgA, IgM, IgD, IgE
  - heavy chain (H)
    - α, γ, δ, ε, μ
  - light chain (L)
    - λ, κ
  - variable region (V)
  - constant region (C)
  - genes
    - V, D, J genes
    - recognition sequences (RS)
    - V(D)J recombinase
    - RAG-1, RAG-2
- Blood groups
  - ABO, Rh
  - hemolytic disease of the newborn (HDN)

HLA system

histocompatibility antigens

HLA-A, HLA-B, HLA-C, HLA-D

class I (HLA-A, HLA-B, HLA-C)

class II (HLA-D) (and subdivisions)

haplotype

cyclosporin

major histocompatibility complex (MHC)

Adenosine deaminase (ADA)

Human immunodeficiency virus (HIV)

T cell immunodeficiency

Severe combined immunodeficiency (SCID)

gene therapy

DiGeorge syndrome

Acquired Immunodeficiency Syndrome (AIDS)

Autoimmunity

sympathetic ophthalmia

rheumatic fever

insulin-dependent diabetes (IDDM)

Processes/Methods

Immune response

shuffling of genes

primary response

antibody-mediated immunity

cell-mediated immunity

secondary response

Antibody diversity

Transfusion

recipients, donors

Hemolytic disease of the newborn (HDN)

Transplant and skin graft rejection

ankylosing spondylitis

rheumatoid arthritis

X-linked agammaglobulinemia (Burton disease)

Concepts

Immunity

recognition of "self" and "nonself"

Antibody-mediated immunity

Immunological memory

vaccination

Antibody diversity

germline theory

somatic mutation theory

recombination theory

Transfusion incompatibilities

Transplantation incompatibilities

Genetics of immunodeficiency

Immunc tolcrance

Solutions to Problems and Discussion Questions

1. V_L = variable region of the light chain
C_H = constant region of the heavy chain
IgG = an immunoglobin class which represents approximately 80% of the antibodies in the blood.
J = genes that specify a portion of the V region which includes a portion of the hypervariable region.
D = a region between V and J in the heavy immunoglobin chain.

The *recombination theory* is supported most heavily by experimental evidence.

2. There would be a possibility of 18 more combinations, 27 total possible.

3. Notice in the *Insights and Solutions* section in this chapter of the text that the number of combinations is determined by a simple multiplication of the number of genes in each class: V X D X J X C Thus in this case the answer would be 10 V X 30 D X 50 J X 3 C = 45,000.

4. Again, notice in the *Insights and Solutions* section in this chapter, the total number of combinations is determined by simple multiplication. In this case, for the heavy chain 5 V X 10 D X 20 J = 1000, and for the light chain 10 V X 100 J = 1000. The final total would be 1000 X 1000 = 10^6.

5. An *antigen* is a substance (often protein) which elicits the formation of antibodies. An *antibody* is a protein which is produced by specific cells of the immune system in response to an antigen.

6. A *helper T cell* is a subtype of T cell which serves as the master switch for the immune system, serving to "turn on" the immune response. *Suppressor T cells* are the "off" switch of the immune system; they stop or slow down the immune response. Both cell types are produced by stem cells in the bone marrow and while immature, migrate to the thymus to mature.

7. In order for a child to have hemolytic disease of the newborn (HDN), the mother must be able to produce antibodies against Rh^+ blood (*DD* or *Dd*). To produce such antibodies, the mother must be Rh^- (*dd*). An HDN child must have Rh^+ blood to react to antibodies against the D antigen. Since a Rh^- mother can contribute only *d*, the father must be either *DD* or *Dd*, and each HDN child must be *Dd*.

8. Individuals with type AB blood do not produce antibodies against the A or B antigens, therefore they can receive blood from individuals with either or both of these antigens. Since type O blood does not have A or B antigens, it can be given to all other blood types (assuming other antigens or factors are compatible). In cases of blood transfusion, the antigens of the donor's blood are significant. Such antigens can react with or stimulate antibody production in the recipient. If the antibodies of the recipient can react with the antigens of the donor, then a blood transfusion match is not achieved.

9. There are four polypeptide chains in each IgG antibody molecule. The are three different types of polypeptide chains represented in the IgG molecule: κ, λ, γ. Each light chain is encoded by at least three gene segments (C, V, and J) while each heavy chain is encoded by at least four gene segments (C, V, D, and J).

10. The *HLA haplotype* is an array of HLA alleles on a given copy of chromosome 6 in humans. We each carry two copies of chromosome 6 and therefore we each have two haplotypes. HLA alleles are closely linked and are each expressed in a codominant fashion.

11. Foreign (donor) tissues which are not genetically "matched" (not from identical twins) stimulate the immune system to produce specific anitbodies and cytotoxic T cells which destroy the transplanted organ. Histocompatibility and ABO antigens are significant in stimulating the rejection response.

12. The chance that another child would have the same set of histocompatibility antigens would be one in four. The family did beat the odds so to speak.

21

Population Genetics

Vocabulary: Organization and Listing of Terms

Historical

Charles Darwin

The Origin of Species 1859

Alfred Russell Wallace

Hardy - Weinberg

Structures and Substances

Population

gene pool

MN blood groups

Biston betularia

Processes/Methods

Gene frequencies

mutation

rates

dwarfism

migration

selection

fitness

selection coefficient

directional

stabilizing

disruptive

random genetic drift

achromatopsia

Symbolism

p, q

$p + q = 1$

$p^2 + 2pq + q^2 = 1$

Sex-linked genes

males = q

females = q^2

Multiple alleles

$p + q + r = 1$

$p^2 + 2pq + 2pr + q^2 + 2qr + r^2 = 1$

Heterozygote frequency

$\sqrt{q^2}$

$p = 1 - q$

$2pq$

Inbreeding and heterosis

assortative

inbreeding

self-fertilization

consanguineous marriages

coefficient of inbreeding

inbreeding depression

hybrid vigor

heterosis (F21.2)

dominance hypothesis

overdominance

Concepts

Population genetics

Gene pool

allelic frequencies (F21.1)

Population

Hardy-Weinberg assumptions

infinitely large

no random genetic drift

random mating

no selection

no mutation

no migration

Genetic equilibrium

genetic variety

Inbreeding and heterosis (F21.2)

F21.1. Diagram of the relationships among populations, individuals, alleles, and allelic frequencies (p, q).

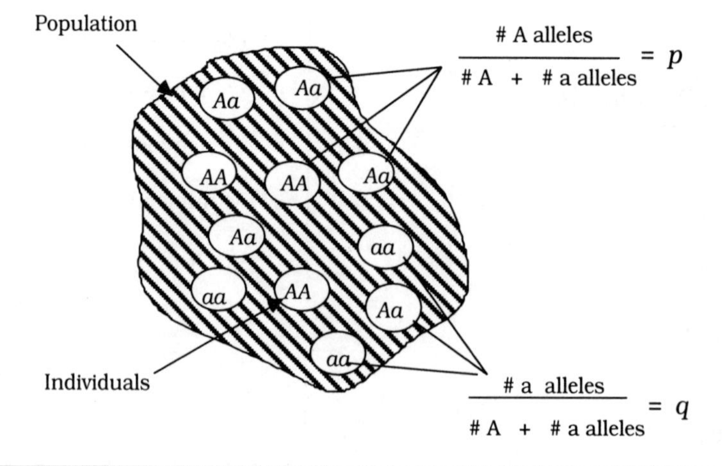

F21.2. Diagram of the relationsips among inbreeding, heterosis, and homozygosity. Note that as inbreeding occurs, heterosis decreases while homozygosity increases.

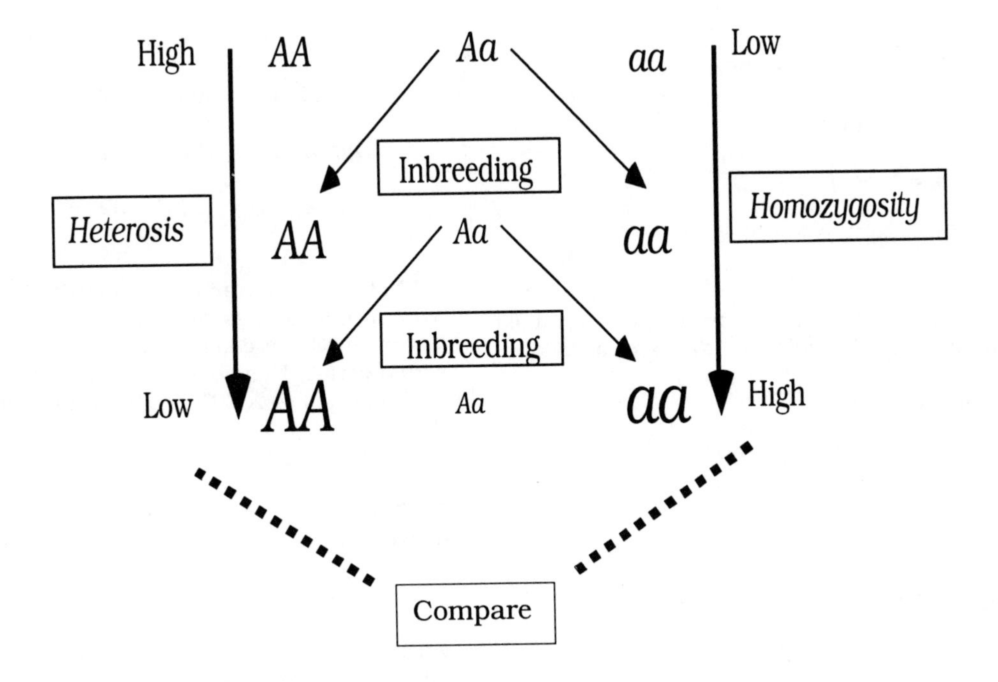

Solutions to Problems and Discussion Questions

1. Because the alleles follow a dominant/recessive mode, one can use the equation $\sqrt{q^2}$ to calculate q from which all other aspects of the answer depend. The frequency of *aa* types is determined by dividing 37 (number of non-tasters) by the total number of individuals (125).

$q^2 = 37/125 = .296$

$q = .544$

$p = 1 - q$

$p = .456$

The frequencies of the genotypes are determined by applying the formula $p^2 + 2pq + q^2$ as follows:

Frequency of *AA* $= p^2$

$= (.456)^2$

$= .208$ or 20.8%

Frequency of *Aa* $= 2pq$

$= 2(.456)(.544)$

$= .496$ or 49.6%

Frequency of *aa* $= q^2$

$= (.544)^2$

$= .296$ or 29.6%

When completing such a set of calculations it is a good practice to add the final percentages to be certain that they total 100%.

2. Given that $q^2 = .04$, then $q = .2$, $2pq = .32$, and $p^2 = .64$ which is the frequency of heterozygotes in the population. Of those not expressing the trait, only a mating between heterozygotes can produce an offspring which expresses the trait, and then only at a frequency of 1/4. The different types of matings possible (those without the trait) in the population, with their frequencies, are given below:

AA X *AA* = .64 X .64 = .4096

AA X *Aa* = .64 X .32 = .2048

Aa X *AA* = .64 X .32 = .2048

Aa X *Aa* = .32 X .32 = .1024

Aa X *Aa* = .32 X .32 = .1024

Notice that of the matings of the individuals who do not express the trait, only the last two (about 20%) are capable of producing offspring with the trait. Therefore one would arrive at a final likelihood of 1/4 X 20% or 5% of the offspring with the trait.

3. The general equation for responding to this question is

$$q_n = q_o/(1 + nq_o)$$

where n = the number of generations, q_o = the initial gene frequency, and q_n = the new gene frequency.

(a)

$q_n = q_o/(1 + nq_o)$

$q_n = 0.5 / [1 + (1 \times 0.5)]$

$q_n = .33$

(b)

$q_n = q_o/(1 + nq_o)$

$q_n = 0.5 / [1 + (5 \times 0.5)]$

$q_n = .143$

(c)

$q_n = q_o/(1 + nq_o)$

$q_n = 0.5 / [1 + (10 \times 0.5)]$

$q_n = .083$

(d)

$q_n = q_o/(1 + nq_o)$

$q_n = 0.5 / [1 + (25 \times 0.5)]$

$q_n = .037$

(e)

$q_n = q_o/(1 + nq_o)$

$q_n = 0.5 / [1 + (100 \times 0.5)]$

$q_n = .0098$

(f)

$q_n = q_o/(1 + nq_o)$

$q_n = 0.5 / [1 + (1000 \times 0.5)]$

$q_n = .00099$

4. (a) The quickest way to generate a homozygous line of an organism is to *self-fertilize* that organism. Because this is not always possible, brother-sister matings are often used.

(b) Notice in the text and in F21.2 that with self-fertilization, the percentage of homozygous individuals increases dramatically.

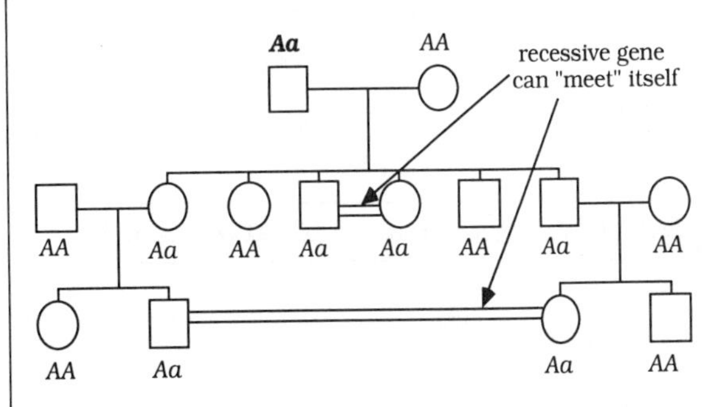

5. The frequency of a gene is determined by a number of factors including the fitness it confers, mutation rate, and input from migration. There is no tendency for a gene to reach any artificial frequency such as 0.5. The distribution of a gene among individuals is determined by mating (population size, inbreeding, etc.) and environmental factors (selection, etc.). A population is in equilibrium when the distribution of genotypes occurs at or around the $p^2 + 2pq + q^2$ expression.

Equilibrium does not mean 25% *AA*, 50% *Aa*, and 25% aa. This confusion often stems from the 1:2:1(or 3:1) ratio seen in Mendelian crosses.

6. The probability that the woman (with no family history of CF) is heterozygous is $2pq$ or 2(1/50)(49/50). The probability that the man is heterozygous is 2/3. The probability that a child with CF will be produced by two heterozygotes is 1/4. Therefore the overall probability of the couple producing a CF child is 98/2500 X 2/3 X 1/4.

22

Genetics and Evolution

Vocabulary: Organization and Listing of Terms

Structures and Substances

- Gene pool
- *Drosophila sp.*
- *Nicotaina sp.*
  - molecular clock
  - divergence dentogram
  - phylogenetic tree
- *Homo sp.*
  - mtDNA

Processes/Methods

- Genetic diversity
  - electrophoresis
  - protein polymorphisms
    - allozymes (F22.1)
  - chromosome polymorphisms
    - inversions, translocations
  - DNA sequence polymorphisms
- Speciation
  - phyletic evolution
  - geographic (allopatric) speciation
    - cladogenesis
  - reproductive isolating mechanisms
    - prezygotic
    - postzygotic
  - race formation
  - sibling species
  - quantum speciation
  - gradualism
  - stochastic
  - catastrophic
  - punctuated equilibrium
  - fonder flush speciation
    - flush, crash
  - allopolyploidy
  - molecular evolution

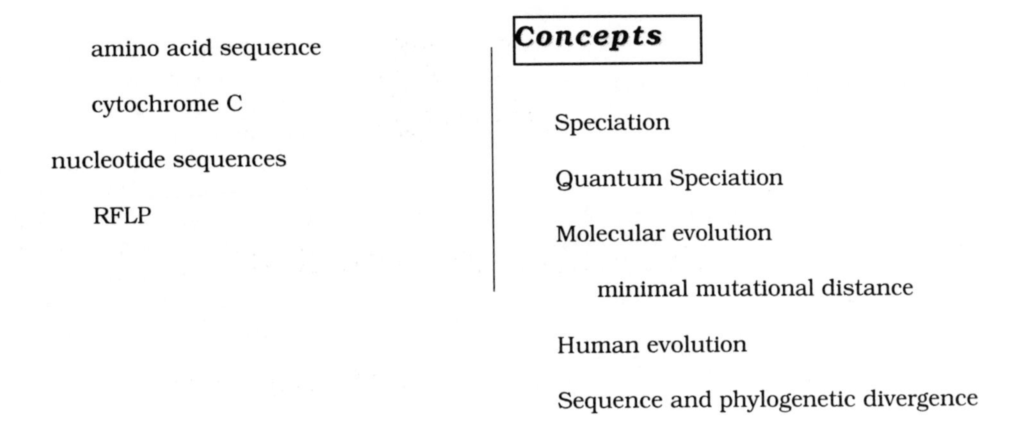

amino acid sequence

cytochrome C

nucleotide sequences

RFLP

Concepts

Speciation

Quantum Speciation

Molecular evolution

minimal mutational distance

Human evolution

Sequence and phylogenetic divergence

F22.1. The diagram below is meant to illustrate the meaning of the term *allozyme*. Notice that alleles, genes *A'* and *A''*, produce protein products which differ electrophoretically but are involved in the same function.

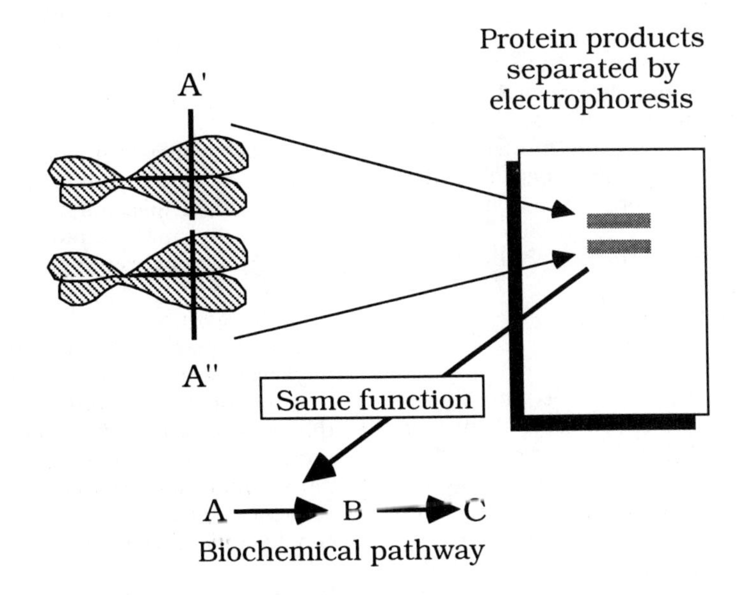

Solutions to Problems and Discussion Questions

1. Assume that a chromosome in the "standard arrangement" undergoes an inversion (pericentric or paracentric). The following are possible consequences of such an inversion:

(a) change in gene order with possible introduction of position effects,

(b) breakage within a structural gene or other functional element,

(c) reduction in the recovery of crossover gametes in heterokaryotypes (those which carry an inversion as well as a standard homologue). While "c" may reduce the production of variation, the first two (a and b) may introduce variation.

2. Approach this problem by writing the possible codons for all the amino acids (except Arg and Asp which show no change) in the human cytochrome c chain. Then determine the minimum number of nucleotide substitutions required for each changed amino acid in the various organisms. Once listed, then count up the numbers for each organism: horse, 3; pig, 2; dog, 3; chicken, 3; bullfrog, 2; fungus, 6.

3. The classification of organisms into different species is based on evidence (morphological, genetic, ecological, etc.) that they are reproductively isolated. That is, there must be evidence that gene flow does not occur among the groups being called different species. Classifications above the species level (genus, family, etc.) are not based on such empirical data. Indeed, classification above the species level is somewhat arbitrary and based on traditions which extend far beyond DNA sequence information.

In addition, recall that DNA sequence divergence is not always directly proportional to morphological, behavioral, or ecological divergence. Therefore, while the genus classifications provided in this problem seem to be invalid, other factors, well beyond simple DNA sequence comparison, must be considered in classification practices. As more information is gained on the meaning of DNA sequence differences (ΔT_m) in comparison to morphological factors, many phylogenetic relationships will be reconsidered and it is possible that adjustments will be needed in some classification schemes.

4. In looking at the figure, notice that the $\Delta T_{50}H$ value of 4.0 on the right could be used as a decision point so that any group which diverged above that line would be considered in the same genus, while any group below would be in a different genus. Under this rule, one would have the chimpanzee, pygmy chimpanzee, human, gorilla and orangutan in the same genus. If one assumed that 3.7 is close enough to be considered above 4.0, given considerable experimental error, one could provide a scheme where the orangutan is not included with the chimpanzee, pygmy chimpanzee, human, and gorilla.

5. Those which occur in DNA, but because of degeneracy in the code, cause no change in the amino acid. In addition, it is possible to exchange a chemically similar amino acid and not detect it in the protein by electrophoresis.

6. DNA sequence data may provide information of phylogenetic divergence; the assumption being that the longer two groups have been separated, the more sequence divergence is expected. However, in coding regions of DNA, selection will narrow the variation and restrict accumulation of sequence changes. Thus it is necessary to qualify conclusions in most cases.

7. Given the small range of *HLA* diversity, one might conclude that the Native Americans descended from a relatively small population either because of a small number who originally arrived or because of significant population crashes over time.

Sample Test Questions (with detailed explanations of answers)

Question 1. Below are phrases which refer to various forms of recombination in bacteria. For each, clearly state whether you **agree** or **disagree**. If you disagree, briefly explain your reason(s).

(a) Transduction is the process in which exogenous DNA is drawn into bacteria as a single-stranded structure, then integrated into the bacterial chromosome.

(b) Temperate phage are capable of entering a lysogenic cycle such that their genomes are incorporated into the bacterial chromosome.

(c) During the lysogenic cycle, phage are capable of producing bacteria when exposed to U.V. light.

Concepts:

recombination in bacteria

transduction

transformation

lysogeny

Answer 1.

(a) Disagree: The process being described refers to *transformation* not transduction. Transduction is *phage-mediated* recombination, whereas in transformation exogenous DNA is taken up as indicated in the statement.

(b) Agree: Viruses which can enter either the lytic or lysogenic cycle are called *temperate* viruses. During the process of lysogeny, the viral chromosome is integrated into the bacterial chromosome as stated.

(c) Disagree: This statement is fairly silly in that it states that phage are capable of producing bacteria. Regardless of the exposure to U.V. light, phage can not produce bacteria. Ultraviolet light can cause induction of the lytic cycle, therefore phage, when lysogenic bacteria are exposed.

Common errors:

carelessness in reading statements

confusion as to what terms mean:

transduction

transformation

conjugation

lysogeny

temperate viruses

Question 2. DNase is often used to map the locations where DNA-binding proteins (histones, RNA polymerases, transcription factors, etc.) interact with DNA. Restriction endonucleases are used to cut DNA for identification and cloning. Why are these two different enzyme classes used in these different ways?

Concepts:

experimental strategies

action of nucleases

DNase

restriction endonucleases

Answer 2. DNase is a general term which includes a variety of exo- and endonucleases which cleave DNA from the ends or internally, respectively. Such cleavage is often irrespective of base sequence. If naked DNA is exposed to DNases, it is rapidly degraded to oligo- and mononucleotides. When protein is associated with DNA, it protects regions from degradation. Such protected regions can be analyzed as to base content. Restriction endonucleases cleave DNA at specific sequences often hundreds or thousands of base pairs apart. If one is interested in mapping protein binding sites, one would want to use an enzyme (a DNase) with frequent yet relatively random cleavage characteristics, not restriction endonucleases.

Common errors:

- **understanding overall strategy**
- **differences between**
  - **DNases**
  - **restriction endonucleases**

Question 3. Assume that you have a cDNA clone for the gene causing retinoblastoma, and you prepare Southern blots probing DNA in cells from normal individuals and from children with retinoblastoma. Genomic DNA is prepared using the restriction endonuclease *Hin*dIII (the *Rb* gene contains four *Hin*dIII fragments as indicated below), and the following hybridization appears:

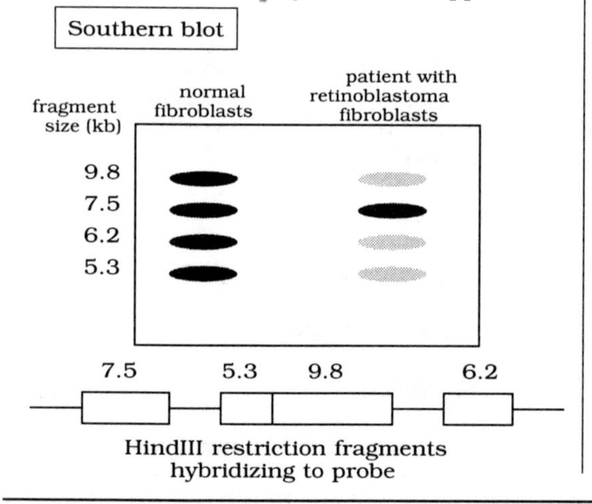

What does each band represent?

What conclusions can be drawn from these data?

Concepts:

- **experimental strategies**
  - **cDNA probes**
  - **Southern blots**
  - **electrophoresis**
  - **restriction endonuclease analysis**
  - **hybridization**

Answer 3. Notice that the fragment sizes (in kb) on the left side of the figure match the *Hin*dIII restriction fragments which are hybridizing (cDNA probe + genomic fragment) to the radioactive probe. Each band therefore represents a region where the radioactive probe is "trapped" by complementary base pairing to single-stranded DNA fragments which are bound to the filter. The smaller fragments migrate faster in the gel and therefore are in the bottom portion while the larger fragments are at the top, near the origin.

Notice that the intensity of the bands from the normal individual is somewhat uniform, indicating that all the restriction fragments are found in equal amounts. However, the intensity of three of the bands from the patient with retinoblastoma are about half as dense as in the normal. One band (7.5 kb) has the same intensity as in the normal.

Because humans are diploid organisms, with normally two copies of each gene, one may hypothesize that the individual with retinoblastoma has a heterozygous deletion of a portion of the retinoblastoma gene which includes *Hin*dIII fragments (9.8, 6.2, and 5.3 kb). It is likely that the inheritance of such a deletion is instrumental in causing familial retinoblastoma.

Common errors:

understanding of experimental design

cDNA probes

Southern blots

electrophoresis

restriction endonuclease analysis

hybridization

recognition of deletion

Question 4. The *thioredoxin* gene in bacteria aids in the necessary reduction of proteins. It encodes a protein of 108 amino acids and is contained in a 0.9 kb (*Pst*I/*Bam*HI) fragment. The gene for kanamycin resistance is contained in a 1.4 kb (*Bam*HI/ *Pst*I) fragment. The restriction map (one orientation) of these two genes (flanked by *Bam*HI sites) in a plasmid vector is presented below.

(A)

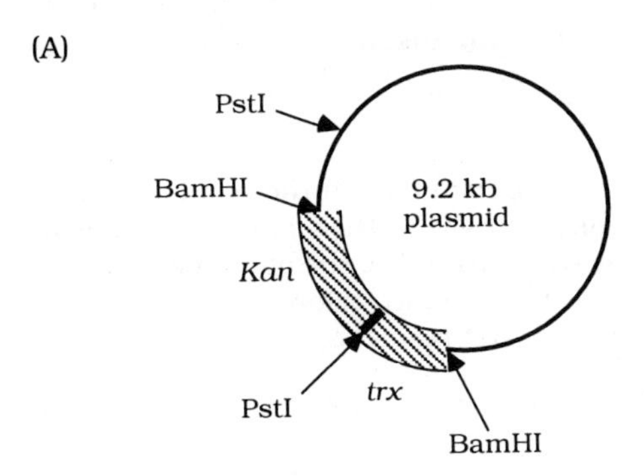

(a) Assume that the plasmid is restricted with the enzyme *Bam*HI. What would be the electrophoretic pattern of the cleaved fragments?

(b) Assume that the orientation given above is only a guess and that the *Bam* HI fragment containing the *trx* and *Kan* genes could possibly exist in the opposite orientation (B). What experiment would you perform to determine whether the orientation is as in (A) or (B)?

(B)

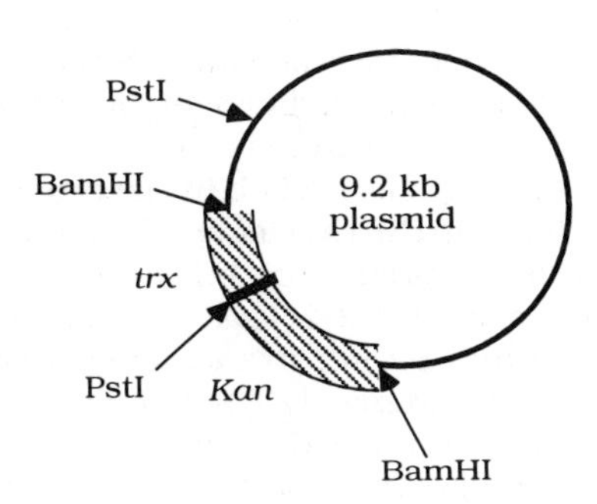

Concepts:

experimental strategies

electrophoresis

restriction digestion

fragment analysis

Answer 4. Below is a drawing of the expected product if either of the above plasmids (A) or (B) is restricted to completion with *Bam*HI. There should be a 2.3 kb fragment (1.4 + 0.9 kb) and the remainder (9.2 - 2.3 = 6.9). Notice that the 6.9 kb fragment migrates slower (higher in the gel) than the 2.3 kb fragment. Also notice that the intensity of the stain is less in the smaller band because there is less DNA to bind the stain.

(a)

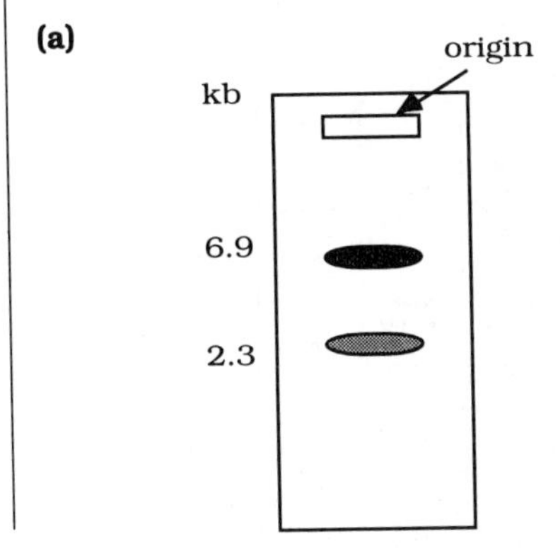

(b) To distinguish between the (A) and (B) orientations, one could make use of the change in position of the *PstI* restriction site in the two orientations. First, estimate the number of kb in the two fragments resulting from *PstI* restriction of orientation (A). Notice that in orientation (A) the *PstI* fragments are approximately 2.4 (1.4 for the *Kan* gene + about 1.0) kb and 6.8 kb. In orientation (B) the sizes would be approximately 1.9 (0.9 for the *trx* gene + about 1.0) and 7.3 kb. With appropriate standards, these size differences could be distinguished on agarose gels.

Common errors:

- **electrophoretic analysis**
- **restriction enzyme analysis**
- **experimental design**

Question 5. The Maxam and Gilbert DNA sequencing procedure involves chemical reactions (methylation) and cleavages (piperidine) that produce ^{32}P-labeled DNA fragments. The chemical reagents can give rise to G, G+A, C, and C+T cleavages. These fragments are then separated on electrophoretic gels, and the sequence is read directly from the gel. A sample gel is given below. From this gel, provide the sequence of the DNA fragment.

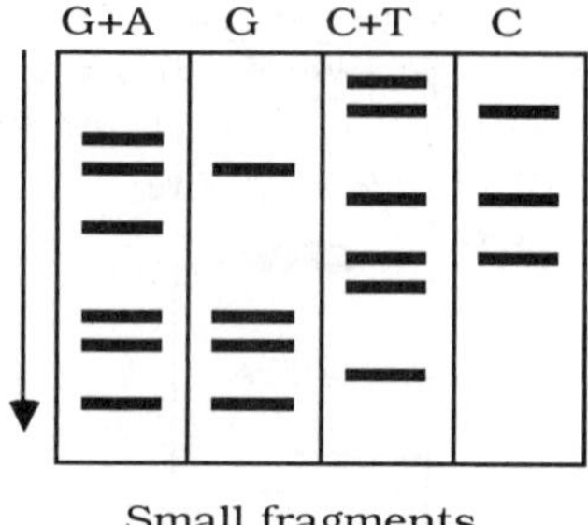

Small fragments at this end of gel

Concepts:

- **electrophoresis**
- **DNA sequencing**

Answer 5. It is a relatively simple procedure to determine the sequence of DNA from the gel given. Start at the bottom with the smallest fragments. As one reads up the gel, one is reading from the 5' to the 3' direction. In the first case, note that there is a band in both the G+A and G lanes. Read this as a "G" because if an "A" occurs, it would exist as a band *only* in the G+A lane. The sequence would be as follows: 5'-GTGGTCACGACT

Common errors:

- **reading the gel as the sequence**
- **difficulty in dealing with G+A and C+T lanes**
- **careless mistakes**

Question 6. Depending on the regulatory system, in prokaryotes when the regulatory protein **is** or **isn't** bound to the DNA, the operon may be **on** or **off**. Fill in the chart below and give a brief explanation of your reasoning.

Relationship of Regulator Protein to DNA	*Operator*	
	Positive control	*Negative control*
is bound		
isn't bound		

Concepts:

genetic regulation in prokaryotes

positive vs. negative control

Answer 6. Refer to F17.1 in this book and notice that any time a regulatory protein interacts with DNA and transcription is stimulated, it is called *positive* control. Any time a regulatory protein interacts with DNA and represses transcription it is called *negative* control. In completing the chart, apply these simple rules.

Relationship of Regulator Protein to DNA	*Operator*	
	Positive control	*Negative control*
is bound	**on**	**off**
isn't bound	**off**	**on**

Common errors:

confusion with positive and negative control

confusion with repressible and inducible systems

Question 7. Acquired Immune Deficiency Syndrome (AIDS) is caused by a retrovirus which exerts its pathogenesis by immunosuppression and selective depletion of helper T lymphocytes. One strategy for AIDS therapy is administration of AZT (azidothymidine) which blocks reverse transcriptase. Present the "life cycle" of a typical retrovirus and show how AZT may act as a therapeutic agent.

Concepts:

retroviral life cycle

action of reverse transcriptase

Answer 7. Retroviruses contain an RNA genome, which in the "life" cycle, is "reverse transcribed" into a DNA strand. First, the single-stranded RNA genome, in the presence of reverse transcriptase, forms a complementary DNA strand. From the DNA strand, a complementary DNA strand is made, thus giving a DNA double-stranded structure which can be integrated into the host DNA. AZT is an inhibitor of reverse transcriptase and therefore blocks a critical step in the cycle mentioned above. Without activity of reverse transcriptase, the RNA genome can not manufacture the complementary strand of DNA.

Common errors:

If students understand the "life" cycle of the virus (retrovirus) there are few difficulties with this type of problem.

Question 8. (a) Describe the general structure of an antibody molecule and the manner in which antibody variability is established.

(b) Assuming that a light chain has 300 V and 10 J regions and the heavy chain has 300 V, 10 D, and 10 J regions, how many different immunoglobulin light-chain genes and how many different heavy-chain genes could theoretically be formed in this organism? Assume that no nucleotides are lost or gained during recombination within the chains.

(c) How many different immunoglobulin molecules could be generated from these heavy and light chains?

Concepts:

immunoglobulin structure

gene recombination

association of chains

Answer 8: (a) An antibody molecule contains two identical light (L) chains and two identical heavy (H) chains. There are two classes of L chains and five classes of H chains. Variability results from variability in the genes encoding the chains and recombination among genes making up each chain.

(b) L: 300 V X 10 J = 3,000, H: 300 V X 10 D X 10 J = 30,000.

(c) Total molecular variation: 3,000 L X 30,000 H = 9 X 10^7

Common errors:

source of variability

general antibody structure

randomness of recombination

careless computation

Question 9. Assume that you have a cDNA clone for the gene causing retinoblastoma, and you prepare Southern blots probing DNA in cells from normal individuals and from children with retinoblastoma. Genomic DNA is prepared using the restriction endonuclease *Hind*III (the *Rb* gene contains four *Hind*III fragments as indicated below), and the following hybridization appears:

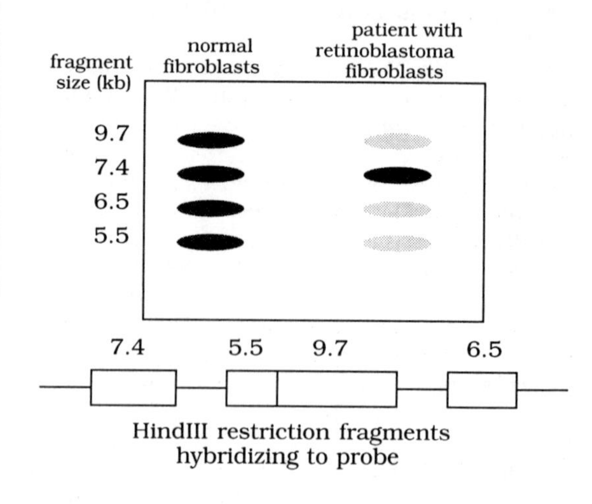

What does each band represent?

What conclusions can be drawn from these data?

Concepts:

experimental strategies

cDNA probes

Southern blots

electrophoresis

restriction endonuclease analysis

hybridization

Answer 9. Notice that the fragment sizes (in kb) on the left side of the figure match the *Hind*III restriction fragments which are hybridizing (cDNA probe + genomic fragment) to the radioactive probe. Each band therefore represents a region where the radioactive probe is "trapped" by complementary base pairing to single-stranded DNA fragments which are bound to the filter. The smaller fragments migrate faster in the gel and therefore are in the bottom portion while the larger fragments are at the top, near the origin.

Notice that the intensity of the bands from the normal individual is somewhat uniform, indicating that all the restriction fragments are found in equal amounts. However, the intensity of three of the bands from the patient with retinoblastoma are about half as dense as in the normal. One band (7.4 kb) has the same intensity as in the normal.

Because humans are diploid organisms, with normally two copies of each gene, one may hypothesize that the individual with retinoblastoma has a heterozygous deletion of a portion of the retinoblastoma gene which includes *Hind*III fragments (9.7, 6.5, and 5.5 kb). It is likely that the inheritance of such a deletion is instrumental in causing familial retinoblastoma.

Common errors:

understanding of experimental design

cDNA probes

Southern blots

electrophoresis

restriction endonuclease analysis

hybridization

recognition of deletion

Question 10. Two terms, *determination* and *differentiation,* are consistently used in discussions of development.

(a) Provide a brief definition of each term.

(b) Which, determination or differentiation, comes first during development of *Drosophila,* for example?

(c) In what way do the two phenomena, *transdetermination* and *regeneration*, provide insights into the processes of determination and differentiation?

Concepts:

terms

determination

differentiation

relationships

transdetermination

regeneration

***Drosophila* development**

Answer 10. (a) *Determination* is a significant, complex, yet poorly understood process whereby the specific pattern of genetic activity is initially established in a cell. This pattern will direct the developmental fate (differentiation) of that cell. *Differentiation* is the process of cellular expression of the determined state. It is the complex series of genetic, morphological, and physiological changes which characterize the variety of adult cells.

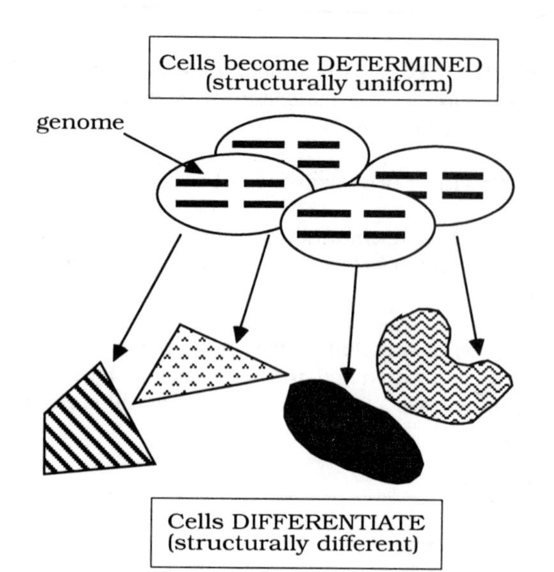

(b) *Determination* occurs before *differentiation*. In *Drosophila*, determinative events are thought to occur about the time of blastoderm formation, when nuclei encounter the peripheral regions of the egg. *Differentiation* of most of the adult cells occurs during metamorphosis, some five to six days after embryogenesis (determination).

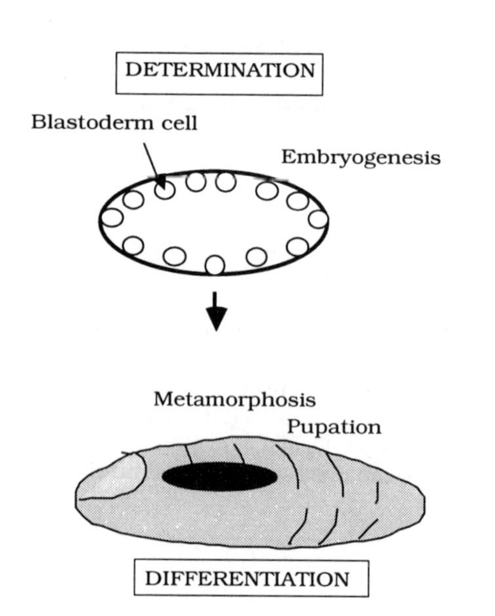

(c) Transdetermination, revealed by the culturing imaginal disks in adult *Drosophila* abdomens, indicates that under some circumstances, determination is not fixed. Thus, the developmental programming of determination may be reversible. Regeneration provides evidence that differentiated cells also lack developmental stability. Cells in the stump portion of an amputation dedifferentiate and accumulate to form a blastema. From that blastema all regenerating tissues will be derived.

Common errors:

providing accurate definitions

failure to relate determination and differentiation to each other temporally and to phenomena of transdetermination and regeneration

Question 11. Development may be defined as the attainment of a differentiated state. Given that all cells of a eukaryote probably contain the same complete set of genes, how do we currently explain development in terms of gene activity? What evidence supports your explanation?

Concepts:

variable gene activity hypothesis

genomic equivalence

evidence for differential transcription

Answer 11. The *variable gene activity hypothesis* of differentiation acknowledges the genomic equivalence of cells within an organism and assumes that of all the genes in a given cell type only certain ones produce products while the others are shut down and are not transcribed. As shown below, certain genes will be active in all cells, those *housekeeping* genes coding for vital cellular functions, while others will be differentially regulated in various cell types. Differential gene transcription occurs in both spatial (different cells of an organism) and temporal (different times during development) dimensions.

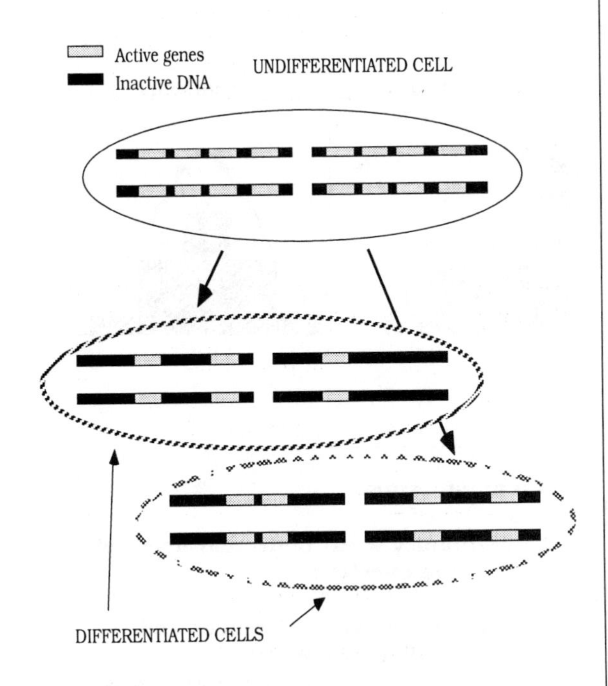

Support for this model is provided by several observations and experiments.

1. *Chromosome puffs:* Specific puff patterns, representing differential gene activity, are observed in dipteran polytene chromosomes at different times during development.

2. *Isozymes:* Differential gene activity is demonstrated by the observation that different forms of the same enzyme (isozymes) are present in cells of different tissues. This evidence assumes that such isozyme patterns are not caused by post-transcriptional forms of genetic regulation.

3. *Growth hormone: In situ* hybridization and immunochemical studies in mouse embryos demonstrates spatial and temporal aspects of the regulation of growth hormone transcripts in the anterior pituitary gland.

Common errors:

With a general question such as this students sometimes have difficulty focusing on the area in their notes or in the text which relates to the question. Students may understand what is meant by the *variable gene activity hypothesis* but not immediately see that it relates to the question. Students also have difficulty in relating a variety of experimental findings to a general theme.

Question 12. Describe similarities and differences between *discontinuous* and *continuous* traits at the *molecular* and *transmission* levels.

> **Concepts:**
>
> **interaction of gene products**
>
> **relationship between genotype and phenotype**
>
> **multi-factor inheritance**

Answer 12. At the *molecular level* one could consider that in discontinuous inheritance the gene products are acting fairly independently of each other, thereby providing a 9:3:3:1 ratio in a dihybrid cross for example. Genotypic classes

A-B-, *A-bb*, *aaB-*, and *aabb*

can be clearly distinguished from each other because the gene products from the *A* locus produce distinct influences on the phenotypes as compared to those gene products from the *B* locus. Exceptions exist where epistasis and other forms of gene interaction occur. In discontinuous inheritance one would consider each locus as providing a *qualitatively* different impact on the phenotype. For instance, even though the *brown* and *scarlet* loci interact in the production of eye pigments in *Drosophila*, each locus is providing qualitatively different input.

In continuous inheritance, we would consider each involved locus as having a quantitative input on the production of a single characteristic of the phenotype. In addition, although it may not always be the case, we would consider each gene product as being qualitatively similar. Under this model, the *quantity* of a particular set of gene products, influenced by a number of gene loci, determines the phenotypic characteristic.

At the *transmission level* one sees "step-wise" distributions in discontinuous inheritance but "smoother" or more bell-shaped distributions in continuous inheritance as shown in the figure below. For instance, in a dihybrid situation (*AaBb* X *AaBb*) where independent assortment holds, one would obtain a 9:3:3:1 ratio (assuming no epistasis, etc.) under a discontinuous mode but a 1:4:6:4:1 where genes (or gene products) are acting additively (continuous inheritance). Both patterns are formed from normal Mendelian principles of segregation, independent assortment, and random union of gametes. It is the manner in which the genes (or gene products) interact which distinguishes discontinuous from continuous inheritance.

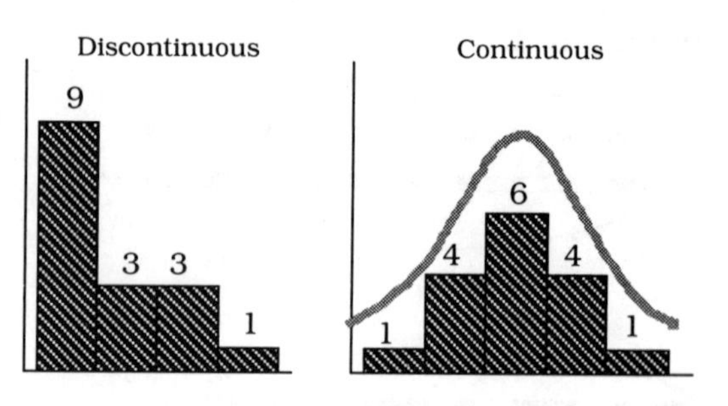

> **Common errors:**
>
> **difficulty with "molecular level" of the question**
>
> **confusion over differences and similarities relating discontinuous and continuous patterns**

Question 13: Phenotypic expression is a complex process involving many levels (including regulatory, metabolic, and developmental) of cellular activity. Two terms, *penetrance* and *expressivity*, are often used to describe phenotypic expression. Provide a definition of each of these terms, then describe factors which vary *penetrance* and *expressivity*.

Concepts:

"global" view of phenotypic expression

relationship of penetrance and expressivity to other factors

Answer 13: *Penetrance* may be defined as the percentage of individuals that express a mutant genotype. Expression may be complete or partial. *Expressivity* refers to the degree of expression of a given genotype. A variety of factors are known to influence penetrance and expressivity. Such factors are expected to act at the levels of gene regulation, transcription, translation, and "higher" developmental processes.

Genetic background. While it is often difficult to assess the influence of genetic background on gene expression, many genes are influenced by "modifiers" which enhance or suppress expression. Since each individual gene functions within an environment produced by all other active genes of the genome, it is reasonable to expect that a host of "background" factors will influence the expression of a gene. Such background factors may involve positional as well as molecular/developmental factors.

Temperature. Since biochemical activity depends on levels of kinetic energy and gene activity and expression are somewhat "biochemical," it is natural to suspect that temperature will influence phenotypic expression.

Nutrition. The raw materials used in growth and maintenance of the living state are usually provided by the diet. As nutrients fluctuate, changes will be expected in the metabolic state of the organism. Such changes may influence the series of biochemical reactions involved in the expression of a gene.

Common errors:

definitions of the penetrance and expressivity

relating the two terms to "environmental factors"

listing and briefly describing the terms

Question 14: Assume that in a particular population, approximately 8% of the males show red-green color blindness. Knowing that this form of color blindness is X-linked, what percentage of the females would be expected to be color blind?

What would be the expected frequency of heterozygous females?

Assuming that the Hardy-Weinberg equilibrium assumptions pertain, what percentage of men will be color blind in the next generation?

Concepts:

Hardy-Weinberg applications to X-linked gene

maintenance of gene frequencies over time

Answer 14: Since 8% of the males express the trait and males have only one X chromosome, the frequency (q) of the recessive gene would be .08 and p would be .92. Since females have two X chromosomes, the expected frequency of females that are homozygous for the color blindness gene would be q^2 or .0064 (.64%).

The frequency of females that are heterozygous would be $2pq$ or 2(.08)(.92) = .1472 or 14.72%.

Because the population is in equilibrium, the frequency of men with color blindness will not change from generation to generation. Eight percent of the men will be color blind in the next generation.

Students should be aware of many deviations of these types of questions. The basic scheme is the Hardy-Weinberg equilibrium and the equations which apply.

Common errors:

application of the Hardy-Weinberg equations to a sex-linked gene

failure to apply the Hardy-Weinberg equations in determining the frequency of heterozygous females

students often make the question harder than it is by forgetting that under equilibrium conditions, gene frequencies do not change

Question 15: List and briefly describe factors which change gene frequencies in populations. Is inbreeding a factor in changing gene frequencies? Explain.

Concepts:

factors which change gene frequencies

influence of inbreeding on gene frequencies

Answer 15:

Mutation, while being an original source of genetic variability, is not usually considered to be a significant factor in changing gene frequencies.

Migration occurs when individuals move from one population to another. The influence of migration on changing gene frequencies is proportional to the differences in gene frequency between the donor and recipient populations. Organisms often migrate as a result of some stress. Those organisms suffering from the most stress are often those that leave. Therefore, they do not represent a random sample of the individuals in that home range.

Selection can be a significant force in changing gene frequencies. It results when some genotypic classes are less likely to produce offspring than others. Selection may be directional, stabilizing, or disruptive.

Genetic drift can be a significant force in changing gene frequencies in populations that are numerically small or have a small number of effective breeders. In such populations, random and relatively large fluctuations in gene frequency occur by "sampling error."

Inbreeding is not a significant factor in changing gene frequencies in populations, however, it will change zygotic or genotypic frequencies. The number of homozygotes will increase at the expense of the heterozygotes.

Common errors:

failure to provide a complete list

failure to briefly and adequately describe each term

failure to see that inbreeding does not, in itself, change gene frequencies

Question 16: A *species* is often defined as a population of interbreeding or potentially interbreeding organisms reproductively isolated from other such populations. Given such an isolated population, will speciation (formation of a new species) occur if the Hardy-Weinberg assumptions are met?

Concept:

relationship between speciation and Hardy-Weinberg assumptions

Answer 16: Reproductive isolation can occur because of the introduction of geographic barriers or other dramatic changes in the environment which subdivide a population. Such factors facilitate speciation because gene flow is eliminated or at least restricted. With gene flow restricted, isolated populations can experience changes in gene frequencies when the Hardy-Weinberg assumptions are *not* met. Under Hardy-Weinberg equilibrium conditions where there is *random mating, no genetic drift, no selection, no mutation,* and *no migration,* gene frequencies will remain the same and speciation will not occur.

Common errors:

confusion as to what the question is asking

difficulty in relating information from one chapter to information contained in a different chapter

Question 17. The study of behavioral genetics is complicated by the fact that many behavioral traits of interest are determined by more than one gene pair and often characterized by variation in penetrance and expressivity. The study of human behavioral genetics suffers additional complications.

(a) List three of these additional complications.

(b) The neurological disorder of Huntington's disease is caused by an autosomal dominant gene, yet in pedigrees it may show incomplete penetrance. Why?

(c) Twin studies have greatly facilitated our understanding of such human behavioral traits as schizophrenia and manic-depression. Why?

(d) What factors are believed to influence the expressivity of human behaviors?

Concepts:

experimental approaches to human genetics

twin studies

penetrance, expressivity

Answer 17:

(a) Several additional problems in the study of human behavioral genetics would include the following.

1. With a relatively small number of offspring produced per mating, standard genetic methods of analysis are difficult.

2. Records on family illnesses, especially behavioral illnesses, are difficult to obtain.

3. The long generation time makes longitudinal (transmission genetics) studies difficult.

4. The scientist can not direct matings that will provide the most informative results.

5. The scientists can't always subject humans to the same types of experimental treatments as other organisms.

(b) With the relatively late age of onset of Huntington's disease, an individual may have children with Huntington's disease, thus indicating the presence of the gene, but die of other causes (accident, military, etc.) before the disease manifests itself. A pedigree having incomplete penetrance will result.

(c) With twin studies, one can readily compare the differential influences of genetic make-up and the environment. Monozygotic twins have the same genetic makeup but can be reared under different environmental conditions if the twins are separated for one reason or another. Dizygotic twins have a different genetic makeup but because they are the same age (almost) they are often reared under fairly uniform conditions within a family. Given these "experimental" advantages, the genetic contribution to certain diseases can often be estimated.

(d) *Genetic background.* While it is often difficult to assess the influence of genetic background on gene expression, many genes are influenced by "modifiers" which enhance or suppress expression. Since each individual gene functions within an environment produced by all other active genes of the genome, it is reasonable to expect that a host of "background" factors will influence the expression of a gene. Such background factors may involve positional as well as molecular/developmental factors.

Environment and *general health.* The physical and behavioral environment in which an individual is raised will influence the degree to which a particular genotype is expressed. Seeing it another way, what may be acceptable or normal behavior in one environment may be abnormal in another. When a living system is stressed by disease or other factors, the chemical/physiological state of the organisms is altered. It is expected that such alterations would be reflected in various behaviors.

Nutrition. The raw materials used in growth and maintenance of the living state are usually provided by the diet. As nutrients fluctuate, changes will be expected in the metabolic state of the organism. Such changes may influence the series of biochemical reactions involved in the expression of a gene. Certain behaviors are known to be influenced by the metabolic state of the individual, thus such behaviors may be influenced by nutrition.

Common errors:

difficulty in relating concepts from different chapters in the text

confusion over penetrance and expressivity

Notes

Notes

Notes

Notes